James E. Bidlack

University of Central Oklahoma

edition fourteen

LABORATORY MANUAL

Stern's Introductory Plant Biology

LABORATORY MANUAL FOR STERN'S INTRODUCTORY PLANT BIOLOGY, FOURTEENTH EDITION

Published by McGraw-Hill, a business unit of The McGraw-Hill Companies, Inc., 1221 Avenue of the Americas, New York, NY 10020. Copyright © 2018 by The McGraw-Hill Companies, Inc. All rights reserved. Printed in the United States of America. Previous editions © 2014, 2011, and 2008. No part of this publication may be reproduced or distributed in any form or by any means, or stored in a database or retrieval system, without the prior written consent of The McGraw-Hill Companies, Inc., including, but not limited to, in any network or other electronic storage or transmission, or broadcast for distance learning.

Some ancillaries, including electronic and print components, may not be available to customers outside the United States.

This book is printed on acid-free paper.

1 2 3 4 5 6 7 8 9 LMN 21 20 19 18 17

ISBN 978-1-260-03014-3
MHID 1-260-03014-8

Chief Product Officer, SVP, Products & Markets: *G. Scott Virkler*
Vice President, General Manager, Products & Markets: *Marty Lange*
Vice President, Content Production & Technology Services: *Betsy Whalen*
Managing Director: *Lynn M. Breithaupt*
Executive Brand Manager: *Michelle Vogler*
Brand Manager: *Justin Wyatt*
Director, Product Development: *Rose Koos*
Executive Marketing Manager: *Patrick E. Reidy*
Product Developer: *Mandy Clark*
Editorial Coordinator: *Jane Peden*
Director of Digital Content: *Michael Koot, PhD*
Program Manager: *Angela FitzPatrick*
Director, Content Design & Delivery: *Linda Meehan-Avenarius*
Content Project Managers: *Mary Jane Lampe/Christina Nelson*
Senior Buyer: *Sandy Ludovissy*
Cover Designer: *Studio Montage*
Cover front and back Image: *Beautiful lupines field over Lake Tekapo: photographer/artist:*
  Ratnakorn Piyasirisorost/Getty Images
Content Licensing Specialists: *Lori Hancock (photo)/Melisa Seegmiller (text)*
Compositor: *SPi Global*
Typeface: *10/12 STIX MathJax Main*
Printer: *LSC Communications—Menasha*

All credits appearing on page or at the end of the book are considered to be an extension of the copyright page.

The Internet addresses listed in the text were accurate at the time of publication. The inclusion of a website does not indicate an endorsement by the authors or McGraw-Hill, and McGraw-Hill does not guarantee the accuracy of the information presented at these sites.

Some of the laboratory experiments included in this text may be hazardous if materials are handled improperly or if procedures are conducted incorrectly. Safety precautions are necessary when you are working with chemicals, glass test tubes, hot water baths, sharp instruments, and the like, or for any procedures that generally require caution. Your school may have set regulations regarding safety procedures that your instructor will explain to you. Should you have any problems with materials or procedures, please ask your instructor for help.

www.mhhe.com

About the Authors

Introductory Plant Biology Laboratory Manual was originally written by Kingsley R. Stern (1927–2006), who spent over 40 years as a devoted botanist and teacher. It has always been Dr. Stern's aspiration that those who use *Stern's Introductory Plant Biology Laboratory Manual* will share his lifelong love of botany. In late 1999/early 2000, Dr. Jim Bidlack joined Kingsley Stern in editing and updating this laboratory manual and has since continued to do so. This fourteenth edition reflects the same accuracy, content, and enthusiasm of the Stern writing style, along with revisions and updates to make it a modern and enjoyable resource for plant biology.

KINGSLEY STERN AND JIM BIDLACK AT KINGSLEY'S OFFICE RESIDENCE IN PARADISE, CALIFORNIA. *Author Image with Kingsley: © Shelley H. Jansky*

Kingsley R. Stern (1927–2006) spent over 40 years as a devoted botanist and teacher. After his formal education (B.S., Wheaton College; M.S., University of Michigan; Ph.D., University of Minnesota), Kingsley educated 15,000 students through classroom/laboratory teaching. He also inspired thousands of readers to appreciate plant biology as author of this laboratory manual and accompanying textbook. Dr. Stern's enthusiasm for the botanical world captivated those around him for many decades. Kingsley Stern will long be remembered for his attention to detail and dedication to high standards, along with a refreshing sense of humor. It was Kingsley's aspiration that those who use this lab manual will share his lifelong love of botany.

James E. Bidlack (B.S., Purdue University; M.S., University of Arkansas; Ph.D., Iowa State University) is a Professor of Biology at the University of Central Oklahoma, where he has spent over 25 years teaching courses in plant biology. He started working with Kingsley Stern in 1998, as a contributor for *Stern's Introductory Plant Biology,* and has since been a coauthor for both the textbook and this lab manual. Like Kingsley, Jim is a devoted botanist and teacher with enthusiasm, high standards, and a sense of humor that students appreciate and enjoy.

Contents

Preface

This laboratory manual assumes no previous knowledge of the biological sciences on the part of the student. It is designed for use in a one-semester or one-quarter introductory course in plant biology and shorter introductory botany courses open to both nonmajors and majors.

Both the principles of biology and the scientific method are introduced, using plants as illustrations. The first few laboratory exercises include introductions to such topics as the microscope, cells, mitosis, roots, stems, and leaves. An exercise on plant propagation, in which students make their own cuttings and grafts, is provided. A functional approach to the topics of cell components and products, diffusion, growth regulation, photosynthesis, and respiration is employed. Exploration of algae, fungi, lichens, bryophytes, ferns, conifers, flowering plants, ecology, and genetics provides students with an understanding of the diversity and complexity of organisms presented in this laboratory manual. Displays of spices, survival plants, and fruit types are suggested. Keys to conifers and woody flowering plants in both their active and dormant conditions are furnished. Each exercise is a unit in itself, allowing for considerable flexibility in sequence of presentation.

If an instructor wishes to use all of the exercises during one semester or quarter, some sessions may include two exercises within the time allotted (e.g., Exercises 2 and 3; Exercises 5 and 6; Exercises 8 and 9; etc.). Materials needed are listed at the beginning of each exercise, and formulae for preparing solutions are listed at the back of the laboratory manual.

Because students observe organisms and structures more closely when they are asked to draw them, some suggestions for student drawings are given at the end of the exercises. In some instances, however, the drawing of an object can become more of a "busy work" exercise than a constructive educational experience. Accordingly, drawings of certain selected organisms and structures are provided for the students to label. Experience suggests this compromise is appropriate for introductory biology laboratory exercises.

The manual is designed so that students can work more or less independently. Instructors are free to require different drawings or other assignments and may also omit some of those suggested within each exercise.

Students are encouraged to read the laboratory exercise before coming to class. To assist them to this end, laboratory preparation quizzes are provided at the end of each exercise. These quizzes should be assigned during the previous laboratory session and collected at the beginning of the class the following week. Answers to the questions are discernible within the particular exercises and should not require checking other sources. In addition, each exercise includes some suggested learning goals and exercise review questions. Answers to the lab manual exercise Review

Questions can be found on the Instructor Edition of the website that accompanies the fourteenth edition of the textbook (http://www.mhhe.com/stern14e).

New to this Edition

The 14th Edition of *Stern's Introductory Plant Biology Laboratory Manual* includes important revisions and updates, most of which focus on photographs and labels for drawings. Several new photographs, including those featured next to each lab number, as well as those in the lab instructions, have been updated to reflect changes in the accompanying textbook. Labels for drawings also have been modified to make them easier to follow, and they generally are listed in a clockwise direction, so they can be easily identified and written in the space provided next to drawings for lab reports. Moreover, keys for labeling diagrams, as well as keys for the review questions and laboratory preparation quizzes, are now available within the Instructor Resources section on Connect™. Some of the more significant revisions include:

- Chapter 1 (The Microscope): The cost for modern electron microscopes has been updated.
- Chapters 3 (Mitosis), 4 (Roots), 5 (Stems), and 6 (Leaves): Labels for drawings have been modified to make them easier to follow.
- Chapter 10 (Photosynthesis): The splitting of water during the light reactions of photosynthesis is explained more accurately.
- Chapter 14 (Domains (Kingdoms) Archaea and Bacteria; Kingdom Protista): Labels for drawings have been modified to make them easier to follow.
- Chapter 15 (Kingdom Fungi (Mycota)): New photographs showing crustose, foliose, and fruticose lichens have been provided and labels for drawings have been modified to make them easier to follow.
- Chapters 16 (Kingdom Plantae: Bryophytes and Ferns), 17 (Kingdom Plantae: Gymnosperms), and 18 (Kingdom Plantae: Angiosperms (Flowering Plants—Phylum Magnoliophyta)): Labels for drawings have been modified to make them easier to follow.

Acknowledgments

Dozens of reviewers have provided input to help revise and update *Stern's Introductory Plant Biology Laboratory Manual*. Contributions and encouragement for this edition were provided by Jan Monelo; colleagues at the University of Central Oklahoma; and the design, editorial, and production staffs of McGraw-Hill Publishers. The author extends special thanks to the following reviewers who provided recent feedback and input used to complete this current edition.

Reviewers who provided feedback for this current edition of the Stern's Introductory Laboratory Manual include:

David Demezas, *University of Wisconsin—Fond du Lac*
Robert Koenig, *Southwest Texas Jr. College*
Rizana Manroof, *South Carolina State University*

Upon reaching this milestone fourteenth edition, thanks are also extended to reviewers of earlier editions, who have provided considerable comments and suggestions:

Ellen Baker, *Santa Monica College*
Rachel Venn Beecham, *Mississippi Valley State University*
Ajoy G. Chakrabarti, *South Carolina State University*
Patricia M. Dooris, *Saint Leo College*
Toye Ekunsanmi, *University of Wisconsin—Washington County*
Joseph Faryniarz, *Naugatuck Valley College*
Scott S. Figdore, *Upper Iowa University*

Christine E. Foley, *Southwest Texas Junior College*
Mort Javadi, *Columbus State Community College*
Dayna S. Lane, *Community College of Baltimore County*
James Leslie, *Adrian College*
Camellia Moses Okpodu, *Norfolk State University*
John Olsen, *Rhodes College*
Murray Paton Pendarvis, *Southeastern Louisiana University*
Thomas Pitzer, *Florida International University*
Linda Mary Reeves, *San Juan College*
Neil Sawyer, *University of Wisconsin—Whitewater*
Nancy Smith-Huerta, *Miami University*
Wendy Stankovich, *University of Wisconsin—Platteville*
Mark A. Storey, *Texarkana College*
John D. Suhr, *Concordia College*
Bibit Traut, *City College of San Francisco*
Roy Turner, *Alvin Community College*
P. Leszek D. Vincent, *University of Missouri, Columbia*

Introduction

A liberally educated person characteristically has at least some familiarity with the scientific method and its applications to everyday living. Indeed, some such knowledge is necessary in order to try to understand oneself and one's interaction with the surrounding environment.

The exercises in this laboratory manual are designed for students who have little or no background in the biological sciences. They provide a working introduction to both the scientific method and the principles of biology, as illustrated by the diversity of plant life around us. The exercises also demonstrate the underlying unity of all living organisms at the cellular level.

Laboratory Materials

You should bring the following materials to each laboratory session: two probes, glass microscope slides, coverslips, single-edge razor blades, drawing paper, a *hard* pencil (3H or harder; note that *hard* refers to the hardness of the lead, *not* to the blackness of the marks produced; hard pencils are preferred because they can be sharpened to a finer point and therefore produce more precise lines), a gum eraser, and a ruler. You can purchase these items separately or as a kit at a bookstore.

Laboratory Drawings and Procedure

Most terms are italicized the first time they are introduced. You are responsible for providing answers to questions that are asked throughout the exercises, for understanding the italicized terms, and for generating the drawings specified by your instructor. You should complete and hand in all laboratory work and drawings during the laboratory period, unless your instructor tells you otherwise. Each week the work will be evaluated on the basis of accuracy, neatness, spelling, and completeness.

It is well known that there is a correlation between closely observing and understanding an organism. The more carefully an object is drawn, the more closely it has been observed. Drawings also are a means of recording facts for future reference. The drawings required in these exercises emphasize careful observation rather than artistic ability. In fact, some common artistic techniques, such as shading or indefinite lines, normally should not be used in biology laboratory drawings, because they may obscure what one is trying to record.

The following rules for laboratory drawings, illustrated in the sample drawing on page ix, should be observed throughout the course and should be periodically reviewed.

1. Draw only on *one* side of the paper, unless your instructor makes an exception to this rule in order to conserve natural resources.
2. *Print* all labels, titles, dates, names, and other information in block capital letters.
3. *Print* your name, laboratory section number, and the date in the upper right corner of each page.
4. Draw each object so that the diameter of the completed drawing is not less than 7.5 centimeters (3 inches).
5. *Print* and center the *title* of the exercise at the top of the page.
6. *Print* a *legend* directly beneath each drawing. Include in the legend the name of the object (e.g., *Ulothrix*), the part being observed (e.g., portion of filament), and the total *magnification* of the drawing (e.g., ×400).
7. Make all lines with a sharp, *hard* pencil. (**Note:** A hard pencil usually has an "H" following the number; a 2H pencil is relatively hard, a 6H pencil is very hard, but a No. 2 or a No. 3 pencil are quite soft.) Do *not* draw any label lines freehand; use a ruler. Keep lines parallel with the edges of the paper, and *never* allow label lines to cross.
8. In labeling, use the name provided in the exercise or texts. Be accurate; if you are labeling a single object, do not use the plural form, and vice versa. Wherever possible, place labels to the right of the drawing, but also use common sense in not applying this rule too rigidly.
9. Allow a blank margin of at least 2.5 centimeters (1 inch) all around the edge of the paper.
10. When making diagrams (line drawings that do not show cellular detail), you may use neat stippling or crosshatching where a greater contrast of two or more adjacent areas is desired, but avoid shading and the use of colored pencils or ink; especially don't make sketchy or otherwise indefinite lines.

NAME _____

LAB SECTION NO. _____

DATE _____

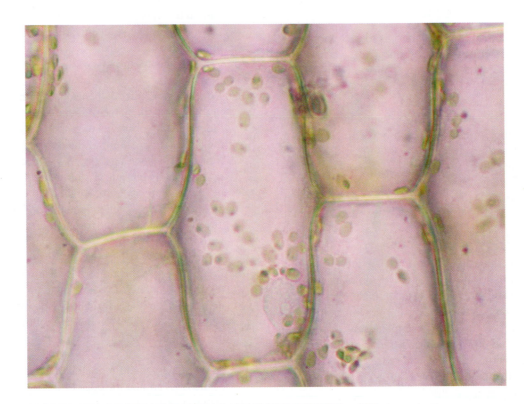

ELODEA CELL AS SEEN THROUGH A LIGHT MICROSCOPE. ×400. © *Kingsley Stern*

Exercise 1

NAME _____

LAB SECTION NO. _____

DATE _____

THE CELL

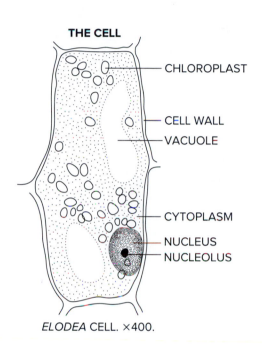

CHLOROPLAST

CELL WALL

VACUOLE

CYTOPLASM

NUCLEUS

NUCLEOLUS

ELODEA CELL. ×400.

The Microscope

Materials

1. Compound microscopes
2. Dissecting microscopes
3. Letter *e* slides
4. Crossed silk fiber slides
5. Pond water

Some Suggested Learning Goals

1. Know the parts of a compound microscope, and understand the function of each part.
2. Understand the differences and similarities between a compound microscope and a dissecting microscope.
3. Be able to calculate the magnification, with specific combinations of lenses, of each object being viewed.
4. Know what is meant by *resolution, field of view,* and *depth of field.*

Introduction

Microscopes were invented at the end of the 16th century and put to use during the second half of the 17th century. Those in use today include variations of two general types.

Small objects are magnified by a system of glass lenses and light with widely used *light microscopes,* such as those available to you in your laboratory. *Electron microscopes* cost more and require more space than light microscopes, but achieve greater magnifications with a beam of electrons controlled by electromagnets. Scientists have discovered an innovative use of electrons in a fairly new type of technology called scanning tunneling microscopy. This technique employs a minute probe that "tunnels" electrons upon a specimen to produce a map of the sample surface. The resolution of images produced by a scanning tunneling microscope is so efficient that even atoms can become discernible.

A. Light Microscopes

General Instructions

Light microscopes are expensive precision instruments that will give satisfactory service for many years if they are handled and maintained properly. The student is responsible for such handling and maintenance while in the laboratory. Read and observe the following instructions carefully:

1. *Always use both hands* when carrying a microscope, and always carry it in an *upright* position. Parts of a microscope may fall off if this rule is not followed.

2. Dirt is the biggest enemy of optical instruments. Keep all parts of your instrument clean. *Use only the lens paper provided to clean the lenses—never use facial tissue or handkerchiefs.* The glass of the lenses is different from window glass; it is relatively soft and scratches easily. It is also easily corroded by acids present in fingerprints and by other chemicals, and rubbing can alter the curvature of the lens. *Do not remove oculars or objective lenses.* When this is done, invisible dust present in the air gets inside the microscope and then may become visible as it is greatly magnified. If dirt or fingerprints are not readily removed with dry lens paper, wet a corner of the lens paper with clean water and *rub gently with a circular motion.* Immediately wipe off anything (including water) that spills on your microscope.

3. Never force any of the adjustments. If something does not work smoothly, call the instructor.

4. Inspect your microscope carefully each time you use it. Report any missing or damaged parts to your instructor *before* you start to use the instrument. *You will be held responsible if you fail to do so.*

5. If you happen to have a compound microscope that can be tilted, use it *only* in the upright position. You gain nothing by tilting, and fluids from slides will run out onto the stage if you do so.

6. Before returning a compound microscope to the microscope cabinet, be sure the *scanning objective* (or *lowest-powered objective*) is in place. If the instrument has a built-in light, wrap the cord around it securely.

B. Compound Microscopes

Parts of a Microscope

Several brands and styles of microscopes may be in use at your institution. Locate the following parts on your compound microscope, and label the photographs of the microscopes at the end of this exercise.

base—This is usually made of heavy metal and is more or less U-shaped. Your microscope has either a lamp or a mirror located between the arms of the U.

arm—This upright metal structure is itself attached to the base and has several other parts attached to it.

barrel or **drawtube**—This is a cylindrical metal tube with a movable *ocular* inserted at the top end.

ocular or **eyepiece**—This is a short metal cylinder with a glass lens toward each end. It may have a hair or line inside that serves as a pointer and moves with the ocular as it is rotated. Most oculars by themselves magnify an object 10 times.

revolving nosepiece—This usually loosely resembles a small, metal pancake sandwich with two to several shotgun shells stuck in it. Note that as you rotate the nosepiece, the metal-cased *objective lenses* must be "clicked" into place for an object to be visible.

objective lenses—There are usually two and often three; in research microscopes, there may be as many as five or six. The magnification achieved by each objective lens is indicated on the side. Common values are 10× for a low-power lens and 40× or 43× for a high-power lens. The total magnification of a viewed object is calculated by multiplying the magnification of the ocular by the magnification of the objective lens. For example, the magnification of an object being viewed with a high-power, 40× objective lens in place would be as follows:

$$10× \text{ (ocular)} × 40× \text{ (h.p. objective)} = 400 \text{ times}$$

stage—This is a platform on which a glass microscope slide to be viewed is placed. The stage has a central hole through which light can pass. The stage may be *mechanical,* with knobs and clamps for moving slides in four directions, or it may have two flexible metal spring clips, each with one end attached to the stage. The spring clips hold a microscope slide in place.

condenser—This is a lens system located beneath the hole in the stage. The condenser may be *fixed* in place, or it may be *adjustable* by means of a knob. It functions in concentrating light in the plane of the object being viewed. If your condenser is adjustable, turn the knob so that the top lens of the condenser is as close to the stage as it will go, then back it off not more than a millimeter, and *leave it there.* At this setting, the maximum amount of light passes through the microscope. Check each time you use your microscope to make sure the condenser is properly adjusted—this is important because some of the resolving power is lost if the condenser is improperly adjusted.

iris diaphragm—This is a series of overlapping, thin metal plates that form a hole through which light can pass. The iris diaphragm is located on or within the condenser; it regulates the amount of light passing through an object. It is adjusted with the little lever protruding from the side of the condenser. You will find you need less light for objects viewed under low power than you will for objects viewed under high power. You will also find that the smaller the hole formed by the iris diaphragm, the greater will be the *depth of focus* (the vertical distance between two adjacent objects, or parts of an object, that are both in focus at the same time). *Always use the iris diaphragm to adjust the amount of light reaching your eye.* **Never** *move the condenser up or down to do this. You can reduce the light intensity by moving the condenser downward, but by doing so you will reduce the resolving power of your microscope.* This is especially important when viewing objects at higher magnifications.

coarse adjustment—This adjustment changes the focus relatively rapidly and is controlled by a knob usually located toward the base of the arm of the microscope. It should be used *only with a low-powered objective lens in place.* Always complete focusing with the *fine adjustment.*

fine adjustment—In some microscopes, this adjustment is brought about by turning a knob separate from the coarse adjustment knob. In other microscopes, the fine adjustment knob may be combined in various ways with the coarse adjustment knob. When there are two distinct knobs, the fine adjustment knob is usually smaller.

lamp or **mirror**—This is located at the base beneath the condenser. The lamp usually has a knoblike switch that should be rotated clockwise to turn it either on or off. The mirror has a flat surface on one side and a concave surface on the other. It should be adjusted to focus the maximum amount of light available on the lower end of the condenser.

Using the Microscope

1. Place the microscope squarely in front of you and just far enough from the edge of the table for you to be able to look through it without becoming uncomfortable. Also, if necessary, adjust the height of your seat.
2. Gently clean all optical surfaces with lens paper.
3. Place the provided letter *e* or crossed silk fibers slide on the stage so that the material is centered over the hole in the middle of the stage. If you have a mechanical stage, center the slide with the stage adjustments; if you have spring clips, place them over the edges of the slide.
4. Switch the *low-power objective* in place, and bring the objective as close as possible to the slide with the coarse adjustment. Always begin examination of a slide with the scanning or low-power objective. The high-power objective is longer than the lower-powered objectives and can crack the slide as well as be damaged if used incorrectly.
5. Turn on the lamp. If you have a mirror, line up the separate lamp provided so that it is about 15 centimeters (6 inches) directly in front of the mirror. Then adjust the mirror so that the maximum amount of light comes through the hole in the stage. Note that the mirror is flat on one surface and concave on the other. In this introductory course, you will obtain your most satisfactory results with the concave surface of the mirror.
6. *Keeping both eyes open,* focus carefully with the coarse adjustment until the material is roughly in focus. If you find you do not see anything with both eyes open, cover one eye with your hand but do not close one eye. Then sharpen the focus with the fine adjustment.

7. Adjust the amount of light reaching your eye with the iris diaphragm. *This is important.* You cannot see detail with light that is either too intense or too dim for your particular eye.

8. Once the material is in focus, move the slide slightly and note that the material moves in the opposite direction. Note also that the image of the material is inverted. Now swing the high-power objective into place and refocus with the fine adjustment. You may need to recenter the material after switching lenses.

9. Completely clean objectives and oculars, clean slides, proper focusing, and correct light intensity are all necessary for maximum *resolution,* which is the capacity of the microscope to separate tiny, closely adjacent objects.

C. Dissecting Microscopes

A dissecting microscope is used for viewing objects that are too thick for light to pass through. It also allows for the examination of much bulkier objects than is possible with a compound microscope. The dual lens system provides a stereoscopic view; magnifications obtained are not as great as those possible with a compound microscope.

1. Compare the construction with that of a compound microscope. Which parts are similar? Which parts are different or absent?

2. Place your hand or some other object on the stage and focus on it. You will probably need to use a separate lamp or some other source of reflected light to view the object clearly.

3. *Interpupillary adjustment.* Because the distance between the eyes varies with individuals, the two lenses are made so that they can be moved together or apart slightly to allow for the variation. The correct distance apart is reached when you can see a single round field.

4. To allow for focus variations in individual eyes, one ocular is also adjustable and can be independently focused. Bring material into focus with the fixed ocular, and then adjust the focus for the other eye by turning the ring at the base of the other ocular. You may need to make this adjustment at each laboratory session, because other sections of the class may also be using your microscope.

5. If you have a "zoom" microscope, there will be a special knob on top of the microscope. Focus material with this knob set at 0.7×. When the material is in focus, you may increase the magnification by turning the knob. If you have a revolving nosepiece, the magnification is changed by rotating the nosepiece. After increasing the magnification, you may find you need to refocus.

D. Electron Microscopes

Electron microscopes (Figs. 1.1 and 1.2) can achieve magnifications of 200,000 times or more and can separate objects as tiny as an angstrom unit (one hundred millionth of a centimeter) wide. They did not come into widespread use until the 1950s. Instead of light, these instruments use a beam of electrons obtained by passing electricity of very high voltage through a V-shaped wire. The wire glows and gives off a stream of electrons that shoots down through a vacuum obtained by pumping air out of a cylindrical column. This column contains electromagnetic lenses spaced along it to control the path of the electron beam. The beam is invisible, but in *transmission electron microscopes* (TEMs) (Fig. 1.1), it is focused on a plate coated with material that fluoresces ("glows") as the electrons strike it. If an object is introduced into the path of the electron beam, the object becomes visible as shadows on the plate. Photographic film responds to the electrons as it would to light, so that clear pictures of objects can be taken by substituting film for the plate.

Preparation of materials for viewing with a TEM involves a number of laborious steps, including embedding the material in plastic, slicing it into extremely thin sections with an *ultramicrotome* (an instrument that operates on the same principle as a meat slicer), and mounting it on copper *grids*. The grids are tiny circular screens about 3 millimeters (0.12 inch) in diameter; they are placed in a holder and introduced into the electron beam in the column for viewing.

Scanning electron microscopes (SEMs) (Fig. 1.2) use the same type of electron source and beam as TEMs, but the beam rapidly scans back and forth over an object on which it is focused. As the electrons bounce off the object's surface, they are picked up by an electronic detector and displayed as an image on a cathode ray tube similar to the picture tube in a television set. A relatively bulky, completely opaque

FIGURE 1.1 A TRANSMISSION ELECTRON MICROSCOPE.
Courtesy of JEOL-USA, Inc., Peabody, MA

FIGURE 1.2   A SCANNING ELECTRON MICROSCOPE.
Courtesy of JEOL-USA, Inc., Peabody, MA

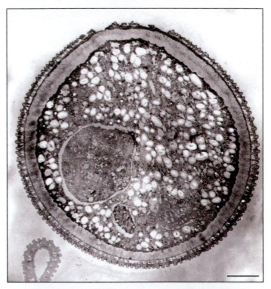

object can be examined by mounting it on a metal stub (usually after the object is coated with a very thin film of metal to heighten contrast). Most SEMs do not have quite as high resolution (capacity to separate tiny objects lying close together) as TEMs, but they are nevertheless far more powerful than any microscopes that use glass lenses and light.

Scanning electron microscopes cost from $75,000 to as much as $1,500,000, whereas transmission electron microscopes start at about $250,000 and can cost as much as $2,000,000. They are powerful instruments and have the ability to reveal some structures that cannot be seen with light microscopy (Fig. 1.3). Electron microscopes are generally much larger than light microscopes and usually are each placed in a room of their own. The expense of these electron microscopes, coupled with extensive sample preparation protocols, restricts their use to advanced courses and research. Accordingly, light microscopy is used in introductory botany courses to provide students with a more economical and straightforward means to learn about plant anatomy and tiny objects.

E. Scanning Tunneling Microscopes

One of the more recent technologies in sample imaging was discovered in 1982 by Gerd Binnig and Heinrich Rohrer, who developed the first scanning tunneling microscope. This instrument uses a minute probe that "tunnels" electrons upon the surface of a sample. The height of the probe is adjusted to keep the flow of electrons constant and fluctuations in probe height are recorded to produce a map of the sample surface. Magnifications achieved with these instruments are so great that the structure of molecules can be revealed. In fact, in 1989, the first picture of a segment of DNA showing its helical structure was produced by an undergraduate student associated with Lawrence Laboratories in northern California.

FIGURE 1.3   POLLEN OF THE WATER LILY (*CABOMBA CAROLINIANA*). *TOP*: A POLLEN GRAIN AS SEEN THROUGH A LIGHT MICROSCOPE (LM). ×1,000. *MIDDLE*: A POLLEN GRAIN AS SEEN THROUGH A TRANSMISSION ELECTRON MICROSCOPE (TEM). ×3,000. *BOTTOM*: A POLLEN GRAIN AS SEEN THROUGH A SCANNING ELECTRON MICROSCOPE (SEM). ×3,000. *Top, middle, and bottom photographs are Courtesy of Jeffrey M. Osborn and Mackenzie L. Taylor.*

Review Questions 1

NAME _____

LAB SECTION NO. _____

DATE _____

1. Why should only lens paper be used to clean microscope lenses? _____

   _____

2. What is the function of the *condenser?* _____

   _____

   What is *resolution?* _____

   _____

   Where should an adjustable condenser be set for maximum resolution? _____

   _____

3. What is the function of the *iris diaphragm?* _____

   _____

   Focus on the letter *e* with the scanning (lowest-power) lens and adjust the iris diaphragm for the correct light intensity. Without adjusting the iris diaphragm further, switch first to low power and then to high power. What happens to the light

   intensity as the magnification is increased? _____

   _____

   While the high-power objective lens is in place, adjust the iris diaphragm to allow the maximum amount of light to be visible. Now switch back to low power and to the scanning lens without adjusting the iris diaphragm. What happens to

   the light intensity? _____

   _____

4. The *field* of vision is the circular lighted area you see when you look through the microscope. What happens to the actual

   amount of field visible (not the size of the objects being viewed) when magnification is increased? _____

   _____

5. Examine the slide labeled "silk fibers," and set it up under low power. Now switch to high power, and *slowly focus away from the slide* with the fine adjustment until the fibers are out of focus. Then *very slowly focus toward the slide* until the first set of fibers comes into focus. These are on top. Continue to focus down *very slowly* until the middle set of fibers comes into focus and finally the bottom set. Record your observations.

Color: top   *Color: middle*   *Color: bottom*   *Your slide letter:*

_____   _____   _____

After you have examined the silk fibers with the high-power lens, examine them with the low-power lens and then the

scanning lens. With which of the three lenses is it easiest to determine the order of the silk fibers? _____

In microscopes, *depth of field* refers to the vertical extent of all the objects in the field of view (the lighted, visible, circular area) that are in focus at the same time. After having used all three lenses in your examination of the silk fibers, you should have noticed a relationship between depth of field and magnification. Complete this sentence to show the

relationship: The greater the magnification, the _____ the depth of field.

6. Calculate the magnification of the lens systems for both dissecting and the compound microscopes.

Dissecting Microscope

Zoom Knob Setting	Ocular Magnification	Total Magnification
0.7		
1.5		
3.0		

Compound Microscope

Lens System	Ocular Magnification	Objective Magnification	Total Magnification
Scanning			
Low power			
High power			

7. What is the function of the *interpupillary adjustment* on the dissecting microscope? _____

_____

8. Label the parts of the microscopes on the illustrations provided on the following pages.

Exercise 1

NAME _____

LAB SECTION NO. _____

DATE _____

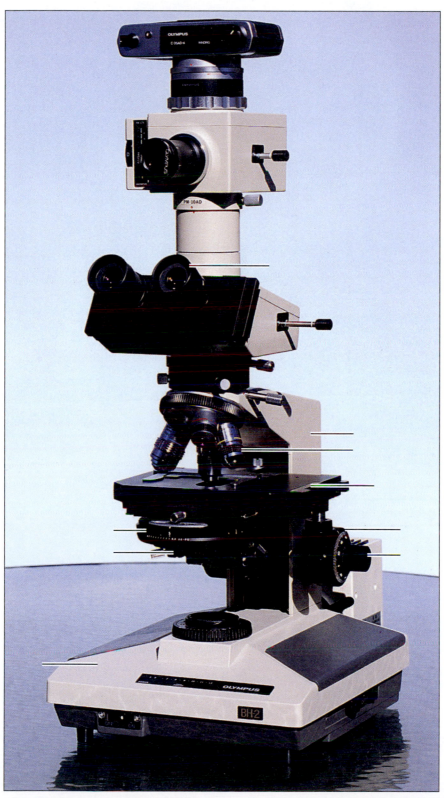

© Kingsley Stern

Exercise 1

NAME _____

LAB SECTION NO. _____

DATE _____

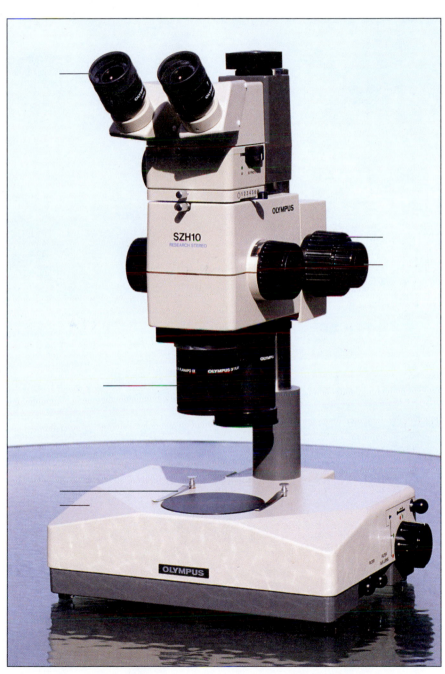

© Kingsley Stern

The Cell

Materials

1. Two or three fresh sprigs of *Elodea*
2. One fresh onion
3. One small, fresh white potato
4. Single-edged razor blades
5. IKI solution (in dropper bottles)
6. Two or three fresh *Tradescantia* flowers (or similar flowers with stamen hairs)
7. Beaker of fresh tap water
8. One ripe tomato or red pepper
9. Bowl of pond water containing various algae and other aquatic organisms
10. Two eyedroppers
11. Plant cell models and charts

Some Suggested Learning Goals

1. Be able to distinguish the various components of living cells visible with a light microscope.
2. Understand the difference between *cyclosis* and independent movement of microscopic objects.

Introduction

Cells are the basic units of which all living organisms are composed. The living material of a cell is called *protoplasm*. Protoplasm is composed mostly of fluid *cytoplasm*, whose primary constituent is water. Cytoplasm also contains a variety of small bodies called *organelles*, membranes, particles, and dissolved substances. The most important organelle is a more or less spherical to ellipsoidal *nucleus* that contains DNA; the nucleus controls the activities of the cell.

Other organelles may include relatively conspicuous green chloroplasts that are often present in cells exposed to light, or *chromoplasts*, which are typically red to orange in color. Other important organelles that are not normally visible with light microscopes include tiny rod- to paddle-shaped *mitochondria*, which function in energy release; *dictyosomes*, which function as packaging centers for substances needed by cells; and *endoplasmic reticulum*, which forms a series of membranous channels connected to the nucleus. Endoplasmic reticulum occurs in a "rough" form, which has tiny, granular-appearing *ribosomes* associated with it, and a "smooth" form without ribosomes. Ribosomes play a role in the manufacture of proteins.

The cytoplasm of a plant cell almost always also includes one or more *vacuoles*. Vacuoles are flexible bags of watery fluid that are bounded by *vacuolar membranes* and that may occupy more than 95% of a plant cell's volume. Vacuolar membranes are similar to the *plasma membrane* that forms the outer boundary of the protoplasm. The plasma membrane is adjacent to the rigid or semirigid *cell wall*, which varies in thickness, depending on the type of cell. Cell walls are visible with a light microscope but vacuolar and plasma membranes are not.

Although cells are produced in a wide variety of sizes and shapes, all have these and other features in common. What features found in living cells are not found in dead cells? After obtaining the various items listed in the "Materials" section, attempt to answer these questions through direct observations, as indicated in the following paragraphs.

A. Cellular Organelles and Cyclosis

Elodea (*Anacharis*) is a widely distributed pondweed that consists of green, submerged stems surrounded by many narrow, flat leaves attached in a tight spiral around the stem. Each leaf is two cells thick, except along the margins, where it is one cell thick. All of the cells are more or less rectangular in outline, but the cells in the upper layer are larger than those in the lower layer, and it is the larger cells we want to examine closely.

Sprigs of *Elodea* are provided. Before obtaining an *Elodea* leaf, be sure that your glass slide and coverslip are completely clean (use detergent to clean them, or alcohol if fingerprints are present). Then, with an eyedropper, add one drop of water from the plant bowl to your slide. The leaves at the tip of an *Elodea* stem may be immature; the older leaves farther down on the stem may have other organisms such as diatoms on their surfaces. Because of this, be sure you remove your leaf from *just a few millimeters below the growing tip*, and also be sure that the upper surface of the leaf you place in the drop of water is facing up (note how the leaf was oriented on the stem). Apply the coverslip to the slide by first dipping one edge in the water drop, and then lower the rest of the coverslip gently until it makes a sandwich of the leaf with the slide. If the leaf is not completely surrounded by water when you have done this, add a little more water at the edge of the coverslip—it will run under on its own.

Using the low-power objective of your microscope, bring the cells of the upper layer into focus. Now switch to high power and refocus. Are all of the cells roughly the same size and shape? Can you detect any movement of the contents of the cells? If at least some of the green *chloroplasts* do not appear to be moving, ask to observe movement on someone else's slide. The movement is called *cyclosis* or *cytoplasmic streaming*. The chloroplasts are not moving under their own power but are being carried along by the riverlike flow of the nearly invisible *cytoplasm*.

Locate a cell with numerous chloroplasts, and focus up and down very carefully and *slowly* with the fine adjustment. Note that at one point all the chloroplasts appear to be only along the margins of the cell. This is because the cells are box-shaped and have depth, even though the leaf may appear to be flat to the unaided eye. Each cell has a large central vacuole bounded by a *vacuolar membrane*. In addition to water, vacuoles sometimes contain pigments or crystals of waste substances. The chloroplasts are located only in the cytoplasm and are plastered up against the six inner walls of the cell, leaving the large, clear vacuole to occupy most of the volume of the cell. When the cell first comes into focus, the chloroplasts appear to be spread across the wall of the cell, which they are. However, further focusing reveals that the cytoplasm is quite thin and confined to the vicinity of the wall, although in some cells, thin strands of cytoplasm, called *cytoplasmic bridges*, may extend across the vacuole.

Locate the thin, semirigid *cell wall* and the *vacuole*. The *nucleus* is often hidden by chloroplasts in *Elodea* cells. If, however, it is visible, it generally appears as a faint, grayish lump about the size of a chloroplast, or a little larger; it is often up against the cell wall. Students who are both patient and determined to see a nucleus are often successful if they first look for a cell that has fewer chloroplasts. To enhance resolution and contrast, be sure to examine the cells with the microscope diaphragm closed so that it admits just enough light to be able to distinguish objects. A *nucleolus*, which appears as a slightly denser spherical body within the nucleus, is often difficult to detect without special staining. The *cytoplasm* is bounded by another invisible membrane, the *plasma membrane*. Also not visible are the *middle lamella*, which is sandwiched between adjacent cell walls, and many smaller organelles present in the cytoplasm. As you move the slide around, you may occasionally observe cells whose vacuoles are pink. The color is due to the presence of water-soluble *anthocyanin pigments*. These pigments are also responsible for some, but not all, of the colors in flowers and ripe fruits.

B. Onion Cells

Your instructor may demonstrate how to peel a single layer of cells from an onion. Mount a segment of the onion peel in a drop of water on a clean microscope slide. Are there chloroplasts present? How many nuclei are present in each cell? To make the nuclei more visible, lift the coverslip, add a drop of IKI (iodine/potassium iodide) solution to the onion segment, and replace the coverslip. This will make the nuclei appear to be golden brown in color (Fig. 2.1).

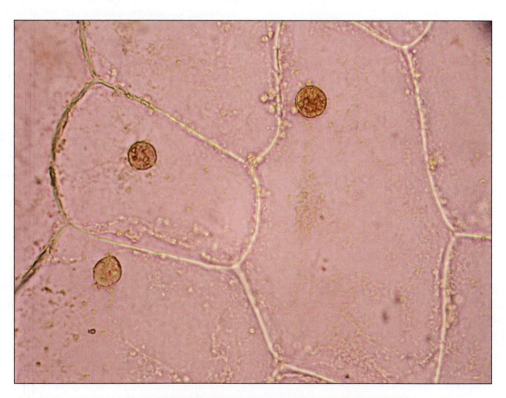

FIGURE 2.1 ONION CELLS, STAINED WITH IKI (IODINE/POTASSIUM IODIDE) SOLUTION, AS SEEN THROUGH A LIGHT MICROSCOPE. ×400. *Courtesy of James E. Bidlack*

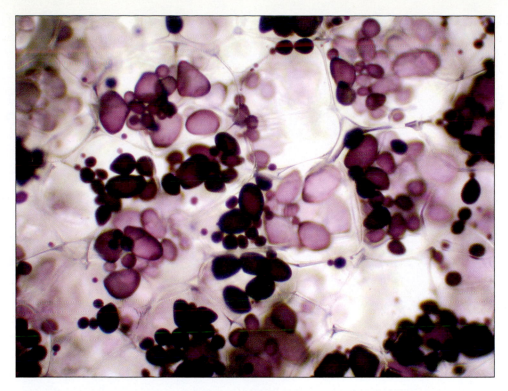

FIGURE 2.2 POTATO PARENCHYMA CELLS, STAINED WITH IKI (IODINE/POTASSIUM IODIDE) SOLUTION, AS SEEN THROUGH A LIGHT MICROSCOPE. ×400. THE DARK OBJECTS ARE STARCH GRAINS FORMED WITHIN COLORLESS AMYLOPLASTS.
Courtesy of James E. Bidlack

C. Potato Parenchyma Cells

The most abundant of plant cells are called *parenchyma* cells. Parenchyma cells occur in various sizes and are thin-walled. The cells usually have several sides (most frequently 14) at maturity. Nearly all the cells of a common white potato are parenchyma cells, which usually contain colorless plastids called *amyloplasts*. These amyloplasts are often clam-shaped in outline and may, when observed under high power, have faint concentric lines formed by deposits of starch within them. Each line represents the limit of one day's deposit of starch. Amyloplasts are quite small at first and increase in size as the starch accumulates.

With a sharp razor blade, make several *paper-thin* sections of potato, and keep the sections wet. Choose the thinnest section, and mount it in a drop of water on a slide. Add a coverslip and, if necessary, another drop of water at the edge of the coverslip so that the whole section is surrounded by water. Locate an area along one edge of the section where the tissue is thin enough for you to distinguish cells. Do not confuse the thin, usually dark and relatively straight cell walls with the numerous starch grains within them (Fig. 2.2). To make the starch grains stand out, add a drop of IKI solution, which stains starch a dark blue-black color, to the edge of your coverslip.

D. *Tradescantia* Stamen Hair Cells

Tradescantia, commonly known as *spiderwort*, produces flowers with pollen-bearing stamens that have many fine hairs on their filaments (stalks). Each hair consists of a single row of connected beadlike cells that become sausage-shaped as they expand.

Have a drop of water ready on a slide, and ask your instructor to give you a *stamen hair* from a *Tradescantia* (spiderwort) or related plant. Cover with a coverslip. If your cells don't have a faint lavender color, they may have been too old or crushed, or they may have been separated from the flower for too long before being immersed in water; if so, you should then obtain another hair. If you focus down carefully under high power, you will note that the surface of each cell is covered with fine *striae*, or lines, which are more or less parallel with each other. Inside the cell, observe the cyclosis occurring, and note that the cytoplasm crisscrosses the large central vacuole via narrow *cytoplasmic bridges*. Also observe the *cytoplasmic granules* in the cytoplasm. Locate the *nucleus*, the *nucleolus*, and the *cell wall*. Are any *chloroplasts* present?

E. Chromoplasts

Ripe tomatoes, red peppers, and several other red to orange fruits owe their color to organelles known as *chromoplasts* within their cells (Fig. 2.3).

Cut a *paper-thin* slice of tissue from the ripe tomato or other red material provided, and mount in water. Locate the small orange or reddish chromoplasts in the cytoplasm. Are the chromoplasts similar in size and shape to *Elodea* chloroplasts? If not, how do they differ?

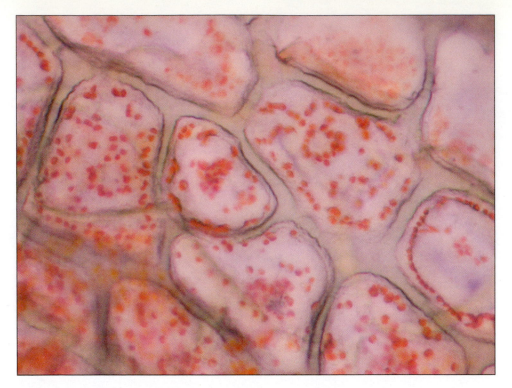

FIGURE 2.3 RED PEPPER CHROMOPLASTS AS SEEN THROUGH A LIGHT MICROSCOPE. ×400. *Courtesy of James E. Bidlack*

F. Pond Water Organisms

Depending on the location, time of the year, and other environmental factors, pond water may contain a rich variety of microscopic living organisms, as well as larger plants and animals.

Stir the provided pond water, and using an eyedropper, place a drop of the agitated water on a slide, cover with a coverslip, and locate as many different kinds of cells or organisms as you can. (Don't confuse shapeless blobs or granules of debris with live cells or organisms.) Your instructor will help you with the names of the algae and other organisms observed. See pages 128–130 for diagrams of some of the cyanobacteria and algae commonly found in pond water.

Drawings to Be Submitted

1. Draw one healthy *Elodea* cell, and label the CELL WALL, CYTOPLASM, VACUOLE, CHLOROPLASTS, and NUCLEUS. If you have observed any other parts of a cell (e.g., NUCLEOLUS), label them also. Remember that each drawing must be made with a *sharp*, hard pencil (2H or harder—*not* an ordinary no. 2 writing pencil) and have a diameter of at least 3 inches. (Review the instructions for laboratory drawings on page v.)
2. Draw one or two potato parenchyma cells. Label the CELL WALL and AMYLOPLASTS.
3. Draw a spiderwort stamen hair cell as viewed under high power. Label the CELL WALL, CYTOPLASM, CYTOPLASMIC BRIDGE(S), NUCLEUS, NUCLEOLUS, CYTOPLASMIC GRANULE(S), STRIAE, and VACUOLE.
4. Draw three different aquatic organisms observed in the agitated pond water.

Review Questions 2

NAME _____

LAB SECTION NO. _____

DATE _____

1. How many layers of cells are there in an *Elodea* leaf? _____

2. How should a coverslip be applied to a drop of liquid on a microscope slide? _____

3. When *chloroplasts* appear to be moving within a living cell, what is the cause of their movement called? _____

    _____

4. In most living cells, such as those of *Elodea*, where is the *cytoplasm* located? How extensive are plant cell *vacuoles?*

    _____

    _____

5. What are *cytoplasmic bridges?* _____

6. What parts of cells are normally visible with the aid of a compound light microscope? _____

    _____

    _____

7. If present in a cell, where are *anthocyanin pigments* located? _____

    _____

8. How are *amyloplasts* distinguished from *parenchyma* cells in a potato? _____

    _____

9. What are *striae*, and where are they located in a spiderwort stamen hair cell? _____

    _____

10. How does a *chromoplast* differ from a *chloroplast?* _____

    _____

Laboratory Preparation Quiz 2

NAME _____

LAB SECTION NO. _____

DATE _____

The Cell

1. In what part of a cell are *chloroplasts* located? _____

2. What is *cyclosis?* _____

3. What is a *vacuole?* _____

   What is the thin boundary of the vacuole called? _____

   Is it visible? _____

4. Where would you look for the *nucleus* in an *Elodea* cell? _____

   _____

5. *Anthocyanin pigments* and *chromoplasts* may both be red in color. If you were to observe a cell that had both, how could

   you distinguish between them? _____

   _____

6. How can you tell a *potato amyloplast* from a *cell?* _____

   _____

7. How would you distinguish an *amyloplast* from a *chloroplast?* _____

   _____

8. What is a *cytoplasmic bridge?* _____

9. Specifically where do *starch grains* develop in a cell? _____

   _____

10. Where would you expect to find a *nucleolus?* _____

Mitosis

Materials

1. Slides of mitosis in onion (*Allium*) root tip
2. Two fresh onions with growing roots
3. One dropper bottle of acetocarmine stain
4. Models and/or charts of cells undergoing mitosis
5. One large sheet of brown paper for each student
6. Modeling clay (two colors)
7. One set of several pipe cleaners (two colors) for each student

Some Suggested Learning Goals

1. Understand and know what takes place in each of the phases of mitosis.
2. Understand the nature, structure, or function of *spindle fibers*, *poles*, *equator*, *chromatids*, *cell plate*, and *phragmoplast*.

Introduction

In *meristematic* tissues (plant tissues in which cells multiply), cells go through an orderly sequence of events known as the *cell cycle*. This cycle is usually divided into *interphase* and *mitosis;* mitosis is typically accompanied by *cytokinesis* (cell division). A clear understanding of the process of mitosis is fundamental to an understanding of growth and reproduction and helps in the comprehension of the important process of meiosis to be studied later. All eukaryotic organisms undergo mitosis in at least some stages of their life cycles. Plant and animal cells differ in some details of the process, but not in the overall scheme. The different parts of the cell cycle can be described as follows:

Interphase—Cells can be easily seen with distinct nuclei, each containing one, two, or more *nucleoli* (which appear as more or less spherical miniature nuclei within a nucleus). Cells in interphase are the most common and should be easy to find. The fine, wispy, lightly stained material is called *chromatin*. During interphase, each chromosome is replicated to form identical copies, called *sister chromatids*. These chromatids are held together at a location called the *centromere*.

Mitosis—This process includes the following four phases, during which the chromosomes separate and two genetically identical daughter nuclei are produced.

prophase—In prophase, the *nuclear envelope* ("casing" of the nucleus) and *nucleoli* disappear, the chromosome strands coil tightly, becoming shorter and thicker, and sister chromatids of each chromosome become distinguishable.

metaphase—During metaphase, each replicated chromosome may resemble the letter "X" with sister chromatids attached at the centromere (Fig. 3.1). The centromeres of chromosomes are aligned at the equator (an invisible plate in the center) of the cell. A top-shaped *spindle*, consisting of numerous *spindle fibers*, becomes fully developed. The spindle fibers extend in arcs between two invisible *poles* located toward opposite ends of the cell. Each chromatid has a single spindle fiber attached to its centromere, but most of the spindle fibers arc directly between the poles.

anaphase—In anaphase, the sister chromatids of each chromosome separate longitudinally at their centromeres. One strand of each pair migrates to a pole, and the other strand of each pair simultaneously migrates to the opposite pole.

telophase—During telophase, the tight coiling of the chromatids—now called chromosomes again—relaxes, and the strands become longer and thinner; a new nuclear envelope develops around each group; and nucleoli reappear. The spindle gradually disappears, and a set of shorter fibers, comprising the *phragmoplast*, develops at right angles to the spindle between the two new nuclei. While this is taking place, dictyosomes in the cytoplasm produce vesicles containing raw materials for cell walls and membranes. Some of these vesicles are channeled to the center of the spindle (equator) by the remaining fibers. As the vesicles accumulate,

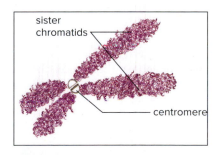

FIGURE 3.1 DIAGRAM OF A DUPLICATED CHROMOSOME SHOWING THE SISTER CHROMATIDS AND CENTROMERE.

they fuse together, forming a double membrane called the *cell plate*. The cell plate grows outward from the center until it contacts and unites with the plasma membrane. Cellulose is then deposited on the membranes, resulting in new cell walls. At the same time, pectin is added between the membranes, creating a *middle lamella* that is shared at the end of telophase by what have become two new *daughter cells*.

A. Cell Cycle in Prepared Microscope Slides

Examine a slide labeled "*Allium* root tip." This is a slide with stained, thin, longitudinal sections (usually three sections) of an onion root tip that was actively growing until it was harvested to make slides. The end of the root is protected by a *root cap* consisting of cells that are irregular in shape. Immediately behind the root cap is a *meristematic region* (*apical meristem*) where cells actively divide. Above the meristematic region, in the *region of elongation*, the cells increase in length without dividing. Many of the cells you will observe will be in interphase because this is the longest phase of the cell cycle.

To find cells in the various stages of mitosis, you will have to confine your examination of the slide to the meristematic region. Starting at the tip of the root, look for cells that demonstrate the different phases of mitosis as well as interphase. Use descriptions in the introduction to identify cells that demonstrate prophase, metaphase, anaphase, and telophase. Once you have located a unique phase, label the different structures observed.

B. Cell Cycle Using Models

Using the large sheets of brown paper provided, draw four boxes, representing cells. Label the first *prophase*, the second *metaphase*, the third *anaphase*, and the fourth *telophase*. Using the modeling clay or pipe cleaners provided, make at least four replicated chromosomes by twisting two lengths of clay or pipe cleaners around each other, to represent how sister chromatids are attached at interphase. The point at which the sister chromatids attach can be viewed as the *centromere* (the dense constricted part of a chromosome to which a spindle filer is attached). If different colors are provided, use the same color for each set of sister chromatids wrapped around each other, and then use a different color for another set of sister chromatids that are the same length. These sets of duplicated chromosomes that

are the same length can be used to represent *homologous chromosomes* (chromosomes that are identical in structure and number of genes, but from different parents). During mitosis, the sister chromatids of each chromosome separate. However, during the first division of *meiosis*, the homologous chromosomes separate, resulting in a reduction in chromosome number.

Note that interphase, which is the longest phase of the cell cycle, can be interpreted as the time when the cell is performing its function as part of a living organism, as well as when the chromosomes duplicate. In this demonstration, interphase may be represented, in part, as what happens when you twist the chromosomes together to prepare for mitotic division. To show the different phases of mitosis, start by placing the duplicated chromosomes randomly on the sheet of paper to demonstrate prophase. Then line up the chromosomes along the equilateral plane of the cell to demonstrate metaphase. Untwist the sister chromatids of each replicated chromosome and begin to separate them to show anaphase. And finally, group the (now unreplicated) chromosomes at opposite poles to view what happens at telophase. Note that the same number of chromosomes, although no longer replicated, exist in each of the two new cells you created.

Optional Exercise

Your instructor may suggest you try to observe the various stages of mitosis in freshly grown onion root tips. If so, place a drop of *acetocarmine* (a mixture of acetic acid and a red stain) provided on a *clean* microscope slide. Then obtain a fresh onion root and remove about 0.5 centimeter of the tapered *tip end*, which should be placed in the acetocarmine drop and allowed to sit for about 5 minutes to soften the tissue. Then with a probe and/or a single-edged razor blade, mash the root tip thoroughly (a second drop of acetocarmine may need to be added). Add a coverslip, and after finding the cells with the low power of your microscope, turn to high power. How many stages of mitosis can you find in the stained cells?

Drawings to Be Submitted

Label the four phases of mitosis (prophase, metaphase, anaphase, telophase) from cells in the meristematic region of an onion root tip. Include the following: NUCLEUS, NUCLEOLUS, NUCLEAR ENVELOPE, CYTOPLASM, SPINDLE FIBERS, CHROMOSOMES, CENTROMERE, CELL WALL, PHRAGMOPLAST, and CELL PLATE. You do not need to label each item more than once.

Exercise 3

NAME _____

LAB SECTION NO. _____

DATE _____

MITOSIS

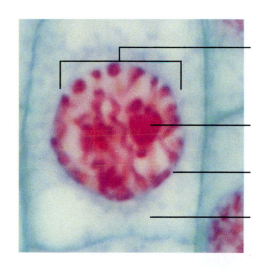

_____

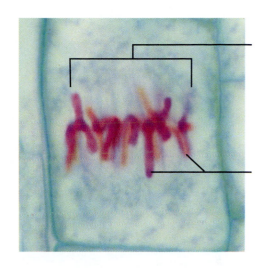

_____

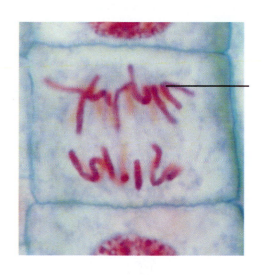

_____

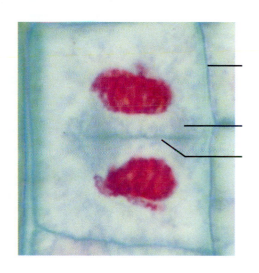

_____

© Kingsley Stern

Review Questions 3

NAME _____

LAB SECTION NO. _____

DATE _____

1. What changes in a cell's cytoplasm and nucleus take place during *interphase?* _____

   _____

2. In which region of a root tip does *mitosis* occur? _____

3. What is a *spindle?* _____

4. What are the main events of *metaphase?* _____

   _____

5. In which phase of mitosis do *chromosomes* first become distinguishable from one another? _____

6. What happens to chromosomes during *anaphase?* _____

   _____

7. Of what substance is a *middle lamella* composed? _____

8. What becomes of the *cell plate* at the conclusion of mitosis? _____

   _____

Laboratory Preparation
Quiz 3

NAME _____

LAB SECTION NO. _____

DATE _____

Mitosis

1. Which plant is used for the study of *mitosis* in this exercise? _____

2. When do single-stranded *chromosomes* become two-stranded? _____

_____

3. In which specific regions of a plant does mitosis commonly take place? _____

_____

4. In which phase of mitosis do *nucleoli* disappear? _____

   When do nucleoli reappear? _____

5. What takes place in *metaphase?* _____

_____

6. Which of the following is visible in at least one stage of mitosis: *poles, equator, spindle?* _____

_____

7. What occurs in *anaphase?* _____

8. From what does a new *cell wall* develop? _____

9. Which cell organelle produces the materials for a new *middle lamella?* _____

_____

10. What is the function of a *centromere?* _____

Roots

Materials

1. Radish or grass seedlings germinated on damp filter paper in petri dishes
2. Prepared slides of cross sections of young buttercup (*Ranunculus*) and greenbrier (*Smilax*) roots
3. Prepared slides of willow (*Salix*) roots showing lateral roots

Some Suggggested Learning Goals

1. Understand the differences between root hairs and lateral roots.
2. Know the locations and functions of root tissues such as *epidermis, cortex, endodermis, pericycle, phloem,* and *xylem.*
3. Know the location and composition of *Casparian strips.*

Introduction

Roots function primarily in anchoring plants and in absorbing water and dissolved substances vital to growth and maintenance of living tissues. The typical regions of a young root tip (*root cap, meristematic region, region of elongation*) were briefly examined in Exercise 3. In this exercise, we want to concentrate on the region behind the region of elongation—the *region of maturation* (also referred to as the *region of differentiation* or *root hair zone*). This region is where the cells originally produced in the meristematic region become differentiated into several different types, each with a specific function.

Flowering plants, primarily on the basis of differences in flower parts, are grouped into two large classes commonly referred to as *dicots* and *monocots*. However, dicots and monocots also differ in the structure of their roots, stems, and leaves. In this exercise, we will briefly examine both dicot and monocot roots.

A. Root Hairs

Root hairs develop in a zone a short distance behind the root cap. As new root hairs are produced near the root cap, the older root hairs farther back die. Root hairs greatly increase the absorptive surface of a root.

Mount about half a centimeter of a living radish or grass seedling root, including the tip, in a drop of water on a slide, add a coverslip, and examine under low power. (Be sure your root tip is intact; if the tip has already been removed, discard the seedling and take another one.) Note the numerous *root hairs* of various sizes. Each root hair is a part of the epidermal cell from which it protrudes; it is *not* a separate cell itself (Fig. 4.1). Can you distinguish the *root cap* that functions primarily in protection of the delicate membranes behind it as the root pushes through the soil?

B. Dicot Roots

Examine a slide showing a cross section of a buttercup root (*Ranunculus* xs). Note that the outermost layer of cells, the *epidermis*, is only one cell thick. Are there any *root hairs* present on your slide? Next note the extensive tissue with numerous *starch grains* (often stained purple) interior to the epidermis. This tissue functions primarily in food storage and is known as the *cortex*. In carrots and similar roots, it comprises the bulk of the root.

The distinctive tissues in the center of the root are surrounded by a single layer of conspicuous cells, most of which appear to have relatively thick walls. This layer, the *endodermis,* forms the inner boundary of the cortex and separates the tissues in the center of the root (known collectively as the *stele*) from the other root tissues. The endodermis is believed to play a role in regulating the movement of water and dissolved substances entering or leaving the stele. Endodermal cells have bands of fatty *suberin* (*Casparian strips*) around the inner faces of the walls. Casparian strips are generally difficult to discern because the fatty substances are dissolved when the slides are being prepared, but in buttercup roots, a conspicuous wall layer (that usually stains red) is deposited inside the Casparian strips. Although suberin itself is impervious to water, endodermal cell walls have many paired *pits* (thin areas where there is no suberin) that allow water to pass through.

The tissue in the center with relatively thick-walled cells (usually stained red) is *primary xylem,* which functions in conducting water. Between the arms of the xylem are patches of *primary phloem,* a food-conducting tissue. In older dicot roots, a *vascular cambium* usually develops between the primary xylem and phloem and produces *secondary xylem* and *phloem.* The addition of secondary tissues by the vascular cambium will increase the girth of the root. Note the *pericycle,* a single layer of thin-walled cells located adjacent to and inside the endodermis. The cells of the pericycle usually do not appear different in form from the young primary phloem cells. *Lateral roots* originate in the pericycle. Unlike dicot stems, dicot roots have no pith.

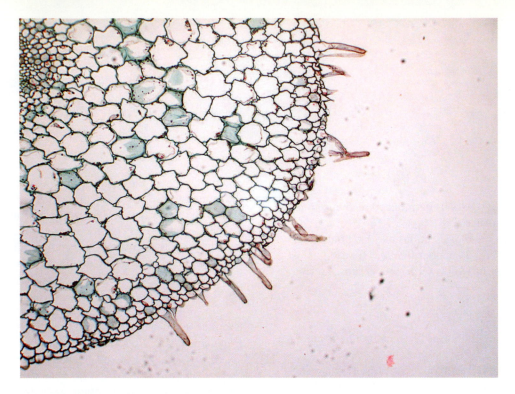

FIGURE 4.1 PARTIAL CROSS SECTION OF A YOUNG PEA (*PISUM SATIVUM*) ROOT SHOWING ROOT HAIRS AND EPIDERMAL CELLS AS SEEN THROUGH A LIGHT MICROSCOPE. ×100. *Courtesy of James E. Bidlack*

C. Monocot Roots

Examine a slide showing a cross section of a root of greenbrier (*Smilax* xs), a monocot. Note the *pith* in the center and the *phloem* within the patches of *xylem*. Locate the *endodermis, pericycle, cortex,* and *epidermis.* What differences and similarities are there between dicot and monocot roots?

D. Lateral Roots

Examine a slide of a cross section of a root of willow (*Salix,* branching xs). This slide shows at least one specially stained lateral root beginning to grow out from the stele. In which specific tissue is the base of the branch root located? Which tissues does it push through as it grows?

E. Specialized Roots and Root Modifications

Your instructor may also provide you with specimens of other types of roots and root modifications for identification. These can include food-storage roots (sweet potatoes and yams), aerial roots (orchids), and prop roots (corn), as well as modifications such as mycorrhizae (most species, particularly visible in pine), and root nodules (most leguminous species). Observe these demonstrations and take note of how these structures assist in the function of roots.

Drawings to Be Submitted

1. Label the provided illustration of a young root through the region of maturation. Indicate the CORTEX, EPIDERMIS, and ROOT HAIRS. Label a ROOT HAIR and an EPIDERMAL CELL on the longitudinal section through a young root, and add a ROOT CAP to the bottom.
2. On the illustration of the cross section of a buttercup root, label EPIDERMIS, CORTEX, STELE, and ENDODERMIS. On the illustration of the enlargement of the stele, label PRIMARY XYLEM, PRIMARY PHLOEM, PERICYCLE, and ENDODERMIS.
3. Label the following on the illustration of the cross section of a greenbrier (*Smilax*) root: EPIDERMIS, CORTEX, ENDODERMIS, PERICYCLE, PRIMARY PHLOEM, PRIMARY XYLEM, and PITH.
4. Label a cross section of a willow root, showing a developing lateral root. Label EPIDERMIS, CORTEX, ENDODERMIS, PERICYCLE, and LATERAL ROOT.

Exercise 4

ROOTS

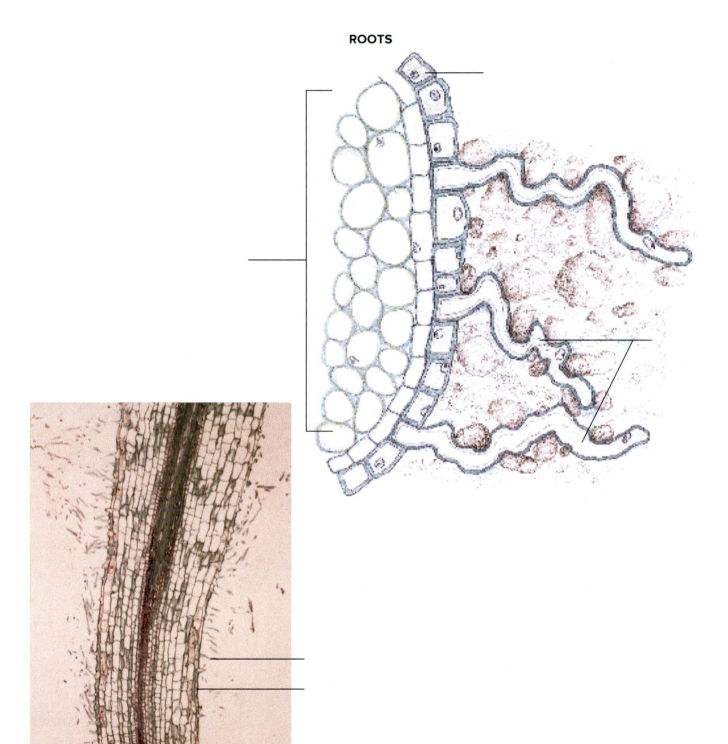

© Kingsley Stern

Exercise 4

ROOTS

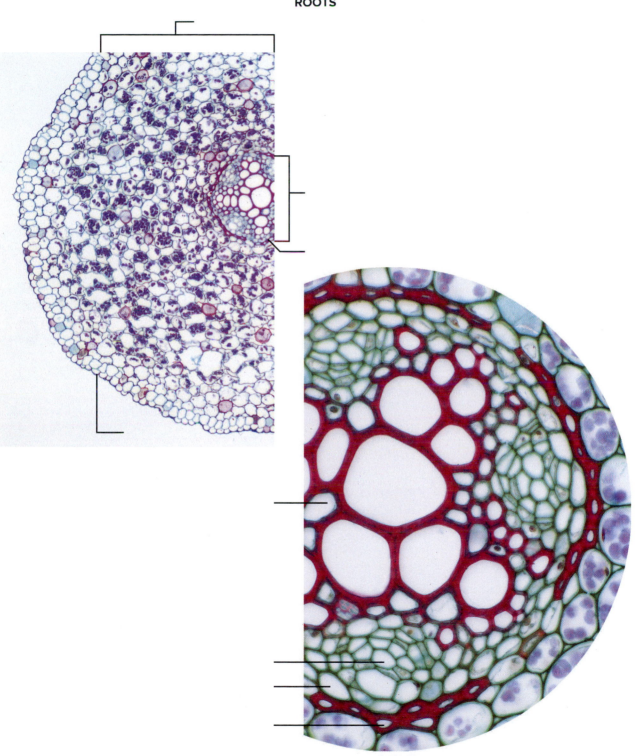

BUTTERCUP (*RANUNCULUS*) ROOT

© Kingsley Stern

Exercise 4

NAME _____

LAB SECTION NO. _____

DATE _____

ROOTS

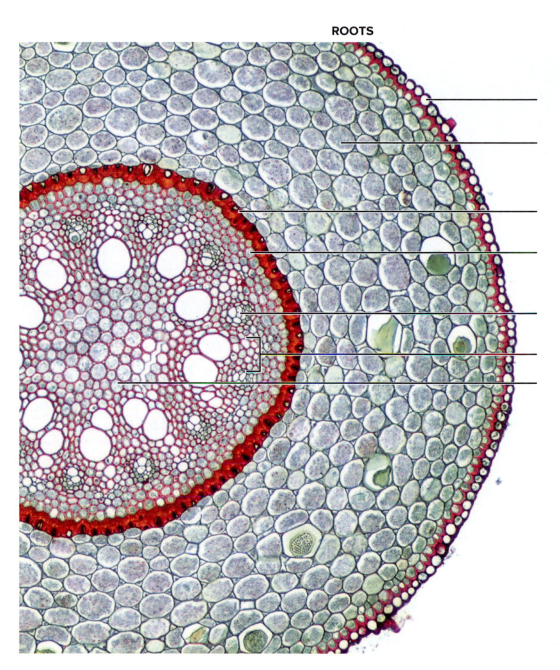

GREENBRIER (*SMILAX*) ROOT

© Kingsley Stern

Exercise 4

NAME _____

LAB SECTION NO. _____

DATE _____

ROOTS

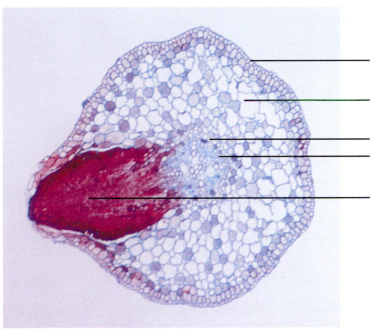

© Kingsley Stern

Review Questions 4

NAME _____

LAB SECTION NO. _____

DATE _____

1. With which specific *region* of roots is this exercise concerned? _____
   _____

2. In which tissues do the following originate? _____

   Root hairs _____ *Lateral roots* _____

3. What evidence of the food-storage function of *cortex* is present in buttercup (*Ranunculus*) roots? _____
   _____

4. Which tissue surrounds and borders the stele of a dicot root? _____

   Which tissues comprise the stele? _____

5. What is the function of the *vascular cambium?* _____
   _____

6. What term is used to describe the bands of fatty substances that are found on the inner surfaces of endodermal cell walls?
   _____

7. Is a *pith* present in all roots? _____ If not, in which roots is it present? _____

8. As lateral roots develop inside a primary root, through which tissues must they grow to reach the surface? _____
   _____

Laboratory Preparation
Quiz 4

NAME _____

LAB SECTION NO. _____

DATE _____

Roots

1. From which tissue do *lateral roots* arise? _____

2. Between which tissues is the *vascular cambium* located? _____

   _____

3. Which tissue of stems is not present in dicot roots? _____

4. In which tissues are *root hairs* to be found? _____

5. Which tissue is immediately adjacent to the *endodermis* on the side toward the center? _____

   _____

6. In which region of the root does differentiation of cells into various cell types take place? _____

   _____

7. What is present in cells of the *cortex* that gives evidence of its function as a food-storage tissue? _____

   _____

8. Of what fatty substance are *Casparian strips* composed? _____

9. What tissue produces cells that add to the girth (diameter) of the root? _____

   _____

10. What water-conducting tissue is present in the center of a dicot root? _____

   _____

Stems

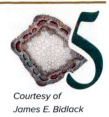

Courtesy of
James E. Bidlack

Materials

1. Dormant twigs of buckeye (*Aesculus*) or similar woody twigs with large buds
2. Prepared slides of cross sections of young and older alfalfa (*Medicago*) stems, 2- to 3-year-old basswood or linden (*Tilia*) stems, and corn (*Zea mays*) stems
3. Live *Begonia* or *Coleus* stems
4. Sharp, single-edged razor blades
5. Live 2-year-old basswood or linden (*Tilia*) twigs
6. Watch glasses containing tap water
7. Sets of seven dropper bottles containing gentian violet stain, eosin stain, phoroglucinol stain, clove oil, xylene, 95% ethyl alcohol, and balsam or *Permount* in xylene
8. Discs of white filter paper (9 centimeters in diameter)
9. Stock bottle of 95% ethyl alcohol
10. Models and charts of both dicot and monocot stems
11. Cake of paraffin

Some Suggested Learning Goals

1. Know the externally visible parts of a woody twig in its winter condition, and know their derivations or functions.
2. Know the locations and functions in a young alfalfa stem of *epidermis, cortex, primary phloem, primary xylem, pith*, and what additional tissues are produced in older alfalfa stems.
3. Understand the differences and similarities between alfalfa and *Begonia* stems.
4. Be able to locate and identify the following in a basswood or linden stem, and know the function of each: *cork, cork cambium, phelloderm, cortex, primary phloem, secondary phloem, vascular cambium, secondary xylem, primary xylem, broad phloem ray, narrow phloem ray, broad xylem ray, narrow xylem ray, fibers, tracheids, vessels, annual ring of xylem*, and *pith*.
5. Know the differences between dicot and monocot stems and the parts and functions of a vascular bundle of corn.

Introduction

Stems and roots are connected to each other and they share most of the same tissues, but the arrangement of the tissues may be different. In addition, a *pith* composed of parenchyma cells is usually seen in the center of dicot stems (pith was absent from the dicot roots examined in Exercise 4). Roots generally have an *endodermis* and a *pericycle*, both of which are usually not seen in stems.

A. External Form of a Woody Twig

Examine the woody dicot twig provided. Note at the tip of the twig the *terminal bud* protected by *bud scales*. Locate a *node* (leaf attachment region of stem), and examine the *leaf scar(s)* present. How many *bundle scars* are present in each leaf scar? (Bundle scars usually appear as tiny bumps where strands of xylem and phloem broke when the leaf fell from the stem; if necessary, use your dissecting microscope to locate them.) Groups of small, narrow scars (*bud scale scars*) often extend around the twig; these sometimes indistinct scars are created when the terminal bud scales of a previous terminal bud fall off. Are any such groups of scars present? If so, how old is your twig?

Note the small, slightly raised *lenticels*, located mostly in the outer bark of internodes (regions of stem between nodes). Lenticels consist of spongy, slightly larger cork cells that may or may not have waxy suberin; they function in permitting exchange of gases (e.g., oxygen, carbon dioxide) between the interior of the stem and the air.

Leaves are usually attached to the stem at an angle. The region immediately above the leaf base and the stem is called an *axil*. Are the *axillary buds* (there is normally one axillary bud in each axil) similar to the terminal buds? If not, how do they differ?

B. Herbaceous Dicot Stems

Select a slide showing a cross section of a young alfalfa stem (*Medicago*, young xs), and examine it under low power. Identify the tissues that have been produced by the apical meristem. These tissues include the *epidermis*, which consists of a single layer of cells that covers all of the exterior of plant organs; the *cortex*, consisting of thin-walled parenchyma cells that function in food storage; *primary xylem*, whose *vessels* and *tracheids* conduct water and minerals; *primary phloem*, whose *sieve-tube members* and associated *companion cells* function in conduction of food in solution; and *pith*, whose parenchyma cells, like those of the cortex, function in food storage. How does the arrangement of the primary xylem and phloem in this stem differ from their arrangement in the buttercup (*Ranunculus*) root you examined in Exercise 4?

Examine an older section of alfalfa stem (*Medicago*, old xs). Note that the xylem and phloem are a little more extensive, due to the addition of *secondary xylem* and *secondary phloem* by the *vascular cambium*, a narrow layer of brick-shaped, meristematic cells that develops between the primary xylem and phloem.

With the aid of a sharp razor blade, cut *paper-thin* cross sections of a *Begonia*, *Coleus*, or other provided stem. Mount in a drop of phoroglucinol (or water) on a slide, and add a coverslip. Examine with the lowest power of your microscope. What similarities between this and alfalfa stems can you detect?

C. Internal Structure of a Woody Stem

Your instructor may have you make your own microscope slide of a woody stem, OR he or she may choose to have you examine previously prepared slides exclusively. If you are to make your own preparation, one procedure is as follows.

Using a sharp razor blade, cut *paper-thin* cross sections of a 2- or 3-year-old basswood or linden (*Tilia*) twig. (**Caution:** The wood is hard; try to brace the twig before cutting so that the razor blade doesn't slip and injure you. You may need to practice a little to get the sections thin enough.) Float the sections in water. Transfer the two thinnest sections, without allowing them to become dry, to a drop of water on a slide. Have *gentian violet stain* ready, and blot off the water with filter paper. Immediately add a drop of the stain to each section. After about 30 seconds, blot off the gentian violet stain, and add a drop of *95% alcohol* to each section. Wait about 1 minute, blot, and add a drop of *eosin stain*. After 1 more minute, add alcohol again, blot off, and add a drop of *clove oil* to each section. The slide should now be ready for the addition of a coverslip and microscopic study.

If you have a good slide and want to preserve it permanently, you may do so by following the clove oil with a drop or two of *xylene*. Then add *balsam*, and place the coverslip on top. After that, the slide will probably take a few days to dry. It can then be stored indefinitely. After you have examined your stained sections, turn to a prepared slide of the same plant (*Tilia* xs) and compare it with the slide(s) you have made.

Contrast the linden stem with the alfalfa and *Begonia* or *Coleus* stems. Note that the phloem in the basswood or linden stem is quite complex and that there are several additional tissues. Focus on the outermost part of the basswood stem first. Note that by the time the stem is 2 or 3 years old, the *epidermis* has been lost and the cells that are sloughing off to the outside are *cork cells*, produced by a *cork cambium* that has developed toward the outer part of the original *cortex*. The cork cambium, consisting of a narrow band of meristematic cells, also produces *phelloderm* cells toward the inside of the stem. The phelloderm cells resemble the cells of the cortex.

To the interior of the cortex is a cylinder of *phloem*, which, as previously mentioned, is quite complex in basswood or linden stems. In cross section, it appears as a circular band of wedges alternating with tapering trapezoids of banded tissue. The wedges include relatively large parenchyma cells that are the flared-out tops of *broad rays*. The rays function in lateral conduction of water, food, and other materials throughout the stem. The part of the ray in the phloem is referred to as a *phloem ray*, while the part of the same ray in the xylem is called a *xylem ray*. In cross section, basswood or linden stems reveal that the broad rays are usually two or three cells wide in the xylem but flare out and become many cells wide in the phloem. *Narrow rays*, however, are usually one cell wide in both the xylem and the phloem. The banded, tapering trapezoids consist of thin-walled *sieve-tube members* and *companion cells* (usually stained green) between bands of thick-walled *fiber* cells that give strength to the stem. The fibers (usually stained red or purple) are often slightly larger at the outside edge of each trapezoid; the larger fibers were produced by the apical meristem as part of the *primary phloem*, while most of the phloem visible in this slide is *secondary phloem* produced by the *vascular cambium*, a narrow band of brick-shaped cells at the base of the wedges and trapezoids.

The vascular cambium also produces *secondary xylem* toward the interior. Note that the xylem (*wood*) appears to have been produced in bands or rings. In fact, each year's production of xylem by the vascular cambium is referred to as an *annual ring*. In basswood or linden stems, each annual ring consists of larger *vessels* that are produced when the cambium first becomes active in the spring, and then progressively smaller vessels and tracheids are produced throughout the remainder of the growing season. Note, again, that the broad phloem rays become broad xylem rays once they cross the vascular cambium and that there are several narrow phloem rays (becoming narrow xylem rays in the xylem), usually one cell wide, between the broad rays. The large, thin-walled parenchyma cells in the center constitute the pith.

D. Monocot Stems

The alfalfa, *Begonia* or *Coleus*, and basswood stems are all *dicot* (dicotyledonous) stems. Examine a slide of corn (*Zea mays* xs) stem, representative of *monocots* (monocotyledonous plants). Note that the xylem and phloem are in *vascular bundles* that are scattered throughout the stem instead of being in a ring as they are in dicot stems. Notice, also, that all the tissues are *primary*; there are no *secondary* tissues because no *cambium* is present to produce them. This also results in there being no separation between cortex and pith, and the parenchyma cells throughout which the vascular bundles are scattered are referred to as *fundamental tissue* instead.

Drawings to Be Submitted

1. Label the provided drawing of the woody twig. The labels should include TERMINAL BUD, BUD SCALE(S), AXILLARY BUD, INTERNODE, NODE, BUD SCAR(S), GIRDLE, and LENTICEL(S).

2. On the illustration of the alfalfa stem provided, label the following: EPIDERMIS, CORTEX, PHLOEM, XYLEM, PITH, and VASCULAR BUNDLE.

3. Label the following on the drawing of a wedge-shaped portion of a cross section of a linden stem provided: BROAD PHLOEM RAY, PRIMARY PHLOEM, NARROW PHLOEM RAY, SECONDARY PHLOEM, BROAD XYLEM RAY, NARROW XYLEM RAY, SECONDARY XYLEM, PRIMARY XYLEM, CORK, CORK CAMBIUM, PHELLODERM, CORTEX, PHLOEM, VASCULAR CAMBIUM, ANNUAL RING OF XYLEM, and PITH.

4. Label the illustration of the monocot (corn) stem provided. Indicate EPIDERMIS, VASCULAR BUNDLE, and FUNDAMENTAL TISSUE on the wedge-shaped portion of the cross section. Label PHLOEM, XYLEM, BUNDLE SHEATH CELL, SIEVE TUBE MEMBER, COMPANION CELL, VESSEL ELEMENT, and TRACHEID on the enlargement of a single vascular bundle.

NAME _____

LAB SECTION NO. _____

DATE _____

STEMS

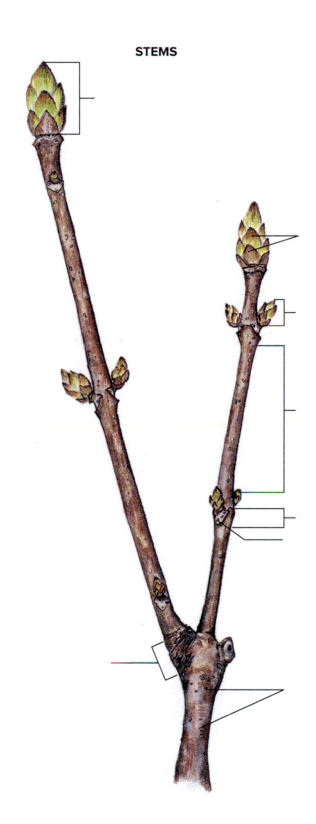

Exercise 5

NAME _____

LAB SECTION NO. _____

DATE _____

STEMS

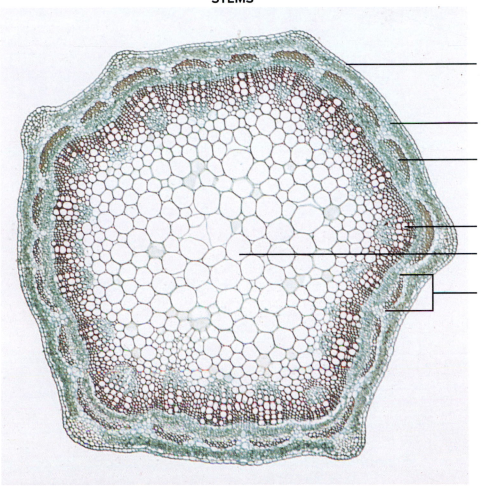

ALFALFA (*MEDICAGO*) STEM

© Kingsley Stern

Exercise 5

NAME _____

LAB SECTION NO. _____

DATE _____

STEMS

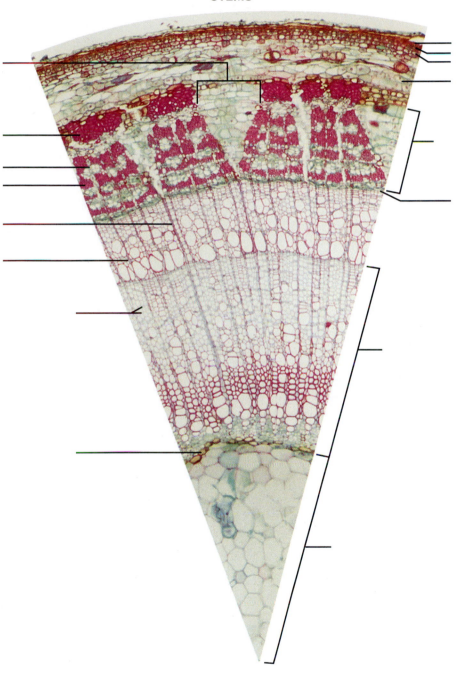

PORTION OF LINDEN (*TILIA*) STEM

© Kingsley Stern

NAME _____

LAB SECTION NO. _____

DATE _____

STEMS

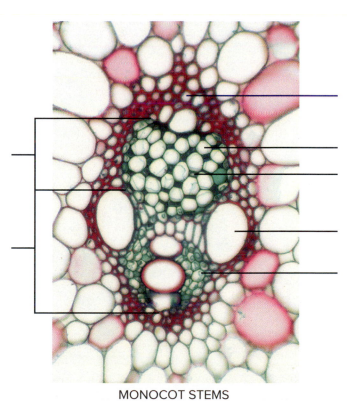

MONOCOT STEMS

© Kingsley Stern

Review Questions 5

NAME _____

LAB SECTION NO. _____

DATE _____

1. What protects the buds of dormant twigs? _____

   _____

2. What are *bundle scars?* _____

   _____

3. Where, specifically, are *axillary buds* located? _____

   _____

4. What structures associated with gas exchange are found throughout stem internodes? _____

   _____

5. What is the difference between *bud scale scars* and *leaf scars?* _____

   _____

   _____

6. Which tissue separates *cortex* from *pith* in an older alfalfa stem? What is the function of this tissue? _____

   _____

7. What is the primary function of *cortex* and *pith?* _____

   _____

8. Which tissue conducts water and minerals in solution? _____

9. If you saw cross sections of *Begonia* or *Coleus* and alfalfa stems side by side, what differences would be obvious? _____

   _____

10. Which stains are used to make the tissues of your handmade linden (basswood) slide more readily visible? _____

    _____

11. If you wished to make your handmade linden (basswood) slide permanent, which additional substances would you use?

    _____

12. Which two tissues are produced by the *cork cambium,* and which two tissues are produced by the *vascular cambium?*

    _____

13. In which tissue(s) of the linden (basswood) stem are *fiber* cells conspicuous? _____

_____

14. Of what kind(s) of cells is an *annual ring* of *xylem* composed? _____

_____

15. What term is applied to *stem parenchyma* tissue that is not separated into cortex and pith? _____

_____

Laboratory Preparation
Quiz 5

NAME _____

LAB SECTION NO. _____

DATE _____

Stems

1. Where are *axillary buds* located? _____

2. What are the small bumps of parenchyma tissue on the surface of the internodes called? _____

   _____

3. How is a *bundle scar* formed? _____

4. What is the function of a *lenticel?* _____

5. Which of the stems in this exercise has the most complex phloem? _____

   _____

6. What stains are used in making your own linden (basswood) slide? _____

   _____

   _____

7. In addition to *cork*, what tissue is usually produced by the *cork cambium?* _____

   _____

8. How are *vascular bundles* arranged in a monocot stem? _____

9. Which of the stems featured in this laboratory exercise is (are) *NOT* (a) dicot(s)? _____

10. To make your own microscope slide of a linden (basswood) stem permanent, what substance would you add just before

    placing a coverslip on it? _____

Leaves

Materials

1. Twigs with simple leaves, pinnately compound leaves, and palmately compound leaves (at least one of the types of leaves should have stipules)
2. Fresh grass (or other monocot) leaves
3. A display of various insectivorous and other modified leaves
4. A healthy *Sedum* or *Zebrina* plant
5. Toothpicks
6. Prepared slides of cross sections of lilac (*Syringa*) and pine (*Pinus*) leaves
7. Prepared slides of hydrophyte versus xerophyte leaves
8. Models and charts of leaves

Some Suggested Learning Goals

1. Understand the difference between a *simple leaf* and a *compound leaf,* and know the parts of a *complete leaf.*
2. Know the differences between the *upper epidermis* and the *lower epidermis* in a lilac (*Syringa*) leaf, and be able to distinguish *guard cells* from other epidermal cells.
3. Be able, with the aid of a compound microscope, to locate *veins (vascular bundles), palisade mesophyll, spongy mesophyll,* and *stomata* in a cross section of a leaf.
4. Know how a pine leaf differs from a dicot leaf with respect to the form and composition of tissues and cells.
5. Understand the distinctions between dicot and monocot leaves.

Introduction

Leaves are important as the principal photosynthetic organs of plants. They are produced in an almost infinite variety of shapes, textures, and sizes. Some of the more common variations are illustrated in the glossary preceding the keys on page 209 and described within the keys. In this exercise, twigs with live leaves are studied first, and common and specialized leaf types are examined with a microscope.

A. Form and Arrangement of Leaves

Examine the provided twigs. Note how the *veins* are arranged in the flattened *blade* of a dicot leaf. Is a *midrib* (larger, central main vein) present? How does the vein arrangement (*venation*) differ from that of a grass (monocot) leaf? Which leaves have *petioles* (stalks)? Are *stipules*

(paired, often leaflike or thornlike appendages at the base of the petiole) present on any of the leaves? Do the *margins* of any of the leaves have sawlike *teeth* or *lobes* (larger, usually rounded projections)? Are hairs or wax present on the surfaces of any of the leaves? Is an *axillary bud* present in the *axil* (the angle—*not* structure—formed by the petiole with the blade) of each leaf? Note that a compound leaf, which is divided into *leaflets,* has a single axillary bud in its axil at the base of the petiole, but there never are buds in the axils of the leaflets. The midrib of a compound leaf is called a *rachis.* A compound leaf with pairs of leaflets arranged along the rachis is said to be *pinnately compound;* a compound leaf whose leaflets fan out from a common point is said to be *palmately compound.*

B. Typical Leaf Structure

Select a slide showing a cross section of a lilac leaf (*Syringa* xs). Note the upper epidermis, which is one cell thick. Normally it is coated with a fatty or waxy *cuticle,* but the cuticle is usually absent in these slides, having been removed by a solvent during the manufacturing process. Immediately below the upper epidermis are two layers of *palisade mesophyll.* Note that the cells are tightly packed together and that they contain numerous chloroplasts. Are chloroplasts present in the upper epidermal cells? Below the palisade mesophyll is the *spongy mesophyll.* Note that the cells of the spongy mesophyll are loosely and somewhat haphazardly arranged. Note also that there are numerous air spaces between them and that spongy mesophyll cells have fewer chloroplasts than palisade mesophyll cells.

Notice that there are *veins (vascular bundles)* scattered throughout the mesophyll. Can you distinguish between thin-walled *phloem* cells in the lower part of a vein and thicker-walled *xylem* cells in the upper part of a vein? Notice also that the veins are of various sizes and that some appear in cross section while others appear to have been sliced at an angle or lengthwise. This is because veins run in various directions and at various angles throughout a lilac leaf blade. Examine the *lower epidermis* that covers the lower or undersurface of the leaf. Do you see any differences between the upper epidermis and the lower epidermis? For one thing, there are *stomata* scattered throughout the lower epidermis. The stomata are formed by pairs of *guard cells.* The guard cells are smaller than the other cells of the lower epidermis, and they may appear slightly recessed. Guard cells, unlike the other epidermal cells, contain chloroplasts that play a role in opening and closing the stomatal pores.

C. Pine Leaf Structure

Turn now to a slide of a pine tree leaf (*Pinus* xs). Pine leaves are adapted to areas where little moisture is available to them when the ground is frozen in the winter, and they look quite different from lilac leaves. Note that the xylem and phloem in the center are surrounded by *transfusion tissue* composed of a mixture of parenchyma cells and short tracheids. The outer boundary of the transfusion tissue is marked by a single row of conspicuous cells comprising the *endodermis*. Notice also, depending on the species of pine, that the xylem and phloem may be in two adjacent patches (*vascular bundles*), or there may be a single vascular bundle. Much of the remaining tissue of the leaf is mesophyll, which is not divided into palisade and spongy layers. Mesophyll is the principal photosynthetic tissue and can be identified by a characteristic rosette shape of individual cells. These rosette cells can facilitate expansion and contraction of pine leaves.

Note the two or more large, circular to elliptical *resin canals* in the mesophyll. The cells lining each resin canal secrete resin into the resin canals. The leaf is covered by an *epidermis*, consisting of a single row of cells, but there are recessed pockets scattered throughout the epidermal cells. Within these pockets, locate the pairs of *guard cells* (they look a little like cats' eyes) that form each *sunken stoma*. Sunken stomata are common in desert plants and in other plants, like pine trees, that grow in areas where moisture may be unavailable or in short supply for at least part of the year. Notice that beneath the epidermis there are one or more layers of thick-walled cells constituting the *hypodermis* (not present in lilac and many other leaves). The hypodermis gives support and rigidity to the pine leaf and also affords a measure of protection to the more delicate tissues of the interior.

D. Stripping and Observing a Leaf Epidermis

Your instructor will show you how to strip the epidermis from a stonecrop, *Zebrina,* or similar leaf. Have a slide with a drop of water ready, and strip a small piece of epidermis for mounting and microscopic examination. Identify the *epidermal cells*, *guard cells*, and *stomata*. Besides the obvious difference in shape, how else do guard cells differ from the surrounding epidermal cells? Are the stomata in your epidermis open or closed?

E. Specialized Leaves

Examine a prepared slide that has cross sections of leaves from desert and aquatic plants. What differences can you see in the *mesophyll*, *epidermis*, and *veins* (*vascular tissue*) of these plants? Explain how the differences adapt the plants to their respective habitats.

Note the display of insect-trapping and other specialized leaves. How do these leaves differ, at least externally, from typical broad leaves of plants of temperate regions?

Drawings to Be Submitted

1. Draw a COMPOUND LEAF attached to a twig. Label BLADE, and where present, PETIOLE, STIPULES, LEAFLET, AXILLARY BUD, and any other visible parts (e.g., HAIRS, SCALES).
2. Fully label the drawing of the stereoscopic view of a portion of a leaf provided. Labels should include VASCULAR BUNDLE (VEIN), STOMA, LOWER EPIDERMIS, UPPER EPIDERMIS, PALISADE MESOPHYLL, SPONGY MESOPHYLL, and GUARD CELLS.
3. Label the portion of the cross section of a lilac (*Syringa*) leaf provided. Labels should include VASCULAR BUNDLE (VEIN), XYLEM, UPPER EPIDERMIS, PALISADE MESOPHYLL, LOWER EPIDERMIS, PHLOEM, STOMA, and SPONGY MESOPHYLL.
4. Diagram and label a cross section of a pine leaf. Labels should include MESOPHYLL, EPIDERMIS, HYPODERMIS, RESIN CANAL, XYLEM, PHLOEM, TRANSFUSION TISSUE, ENDODERMIS, SUNKEN STOMA, and VASCULAR BUNDLE.
5. Draw a portion of the epidermis of a stonecrop leaf, showing at least one STOMA, GUARD CELLS, and surrounding EPIDERMAL CELLS.

NAME _____

LAB SECTION NO. _____

DATE _____

LEAVES

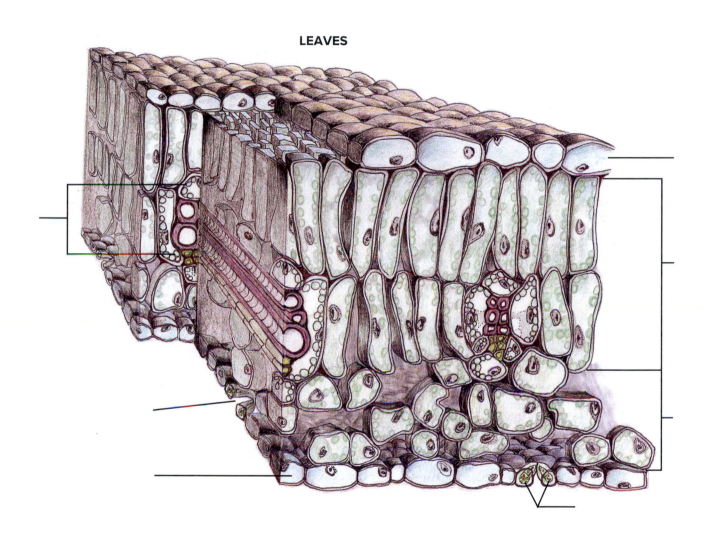

NAME _____

LAB SECTION NO. _____

DATE _____

LEAVES

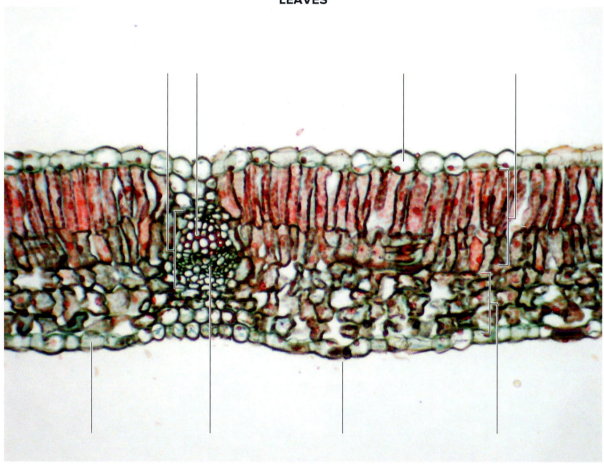

CROSS SECTION OF A LILAC (*SYRINGA*) LEAF

Exercise 6

NAME _____

LAB SECTION NO. _____

DATE _____

LEAVES

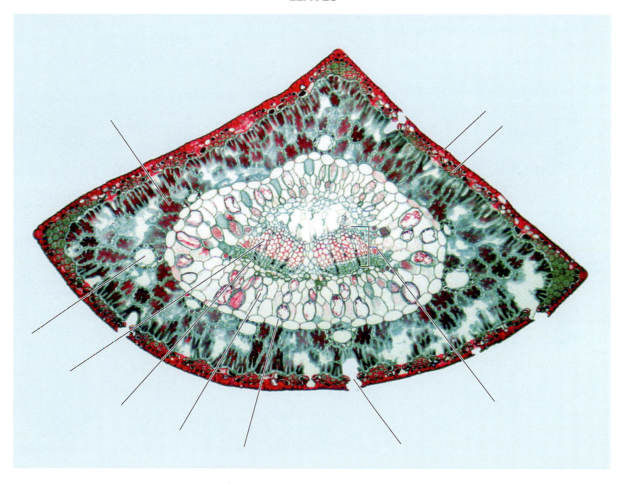

CROSS SECTION OF A PINE (*PINUS*) LEAF

Courtesy of James E. Bidlack

Review Questions 6

NAME _____

LAB SECTION NO. _____

DATE _____

1. How does a *compound* leaf differ from a *simple* leaf? _____

   _____

2. What fatty or waxy substance present on the outer walls of leaf epidermal cells is often lost in the preparation of slides?

   _____

3. When you view a cross section of a leaf with the upper epidermis at the top, where is the *phloem* located in a *vein?*

   _____

4. Which of the larger organelles are most abundant in *palisade mesophyll* cells? _____

5. What specific tissue marks the outer boundary of *transfusion tissue* in a pine leaf? _____

   _____

6. Which tissue lies between the *epidermis* and the *endodermis* in a pine leaf? _____

   _____

7. Where are the *resin canals* located in a pine leaf? _____

   What is their function? _____

8. What are *sunken stomata?* _____

   With which types of plants are they associated? _____

9. What is the function of a *hypodermis?* _____

   Where is a hypodermis located? _____

10. Apart from size and shape, how do *guard cells* differ from the epidermal cells that surround them? _____

    _____

Laboratory Preparation
Quiz 6

NAME _____

LAB SECTION NO. _____

DATE _____

Leaves

1. What are *stipules?* _____

2. What is the fatty or waxy substance that coats a leaf epidermis called? _____

   _____

3. What tissue composed of thick-walled cells is found just beneath the epidermis of a pine leaf? _____

   _____

4. In prepared slides of lilac leaves, why are some *veins* visible in cross section while others are visible in longitudinal

   section? _____

5. Which tissue of pine leaves differs from that of lilac leaves in its not being divided into two distinguishable layers?

   _____

6. Of which two tissues are leaf *veins* primarily composed? _____

7. Where are *stomata* generally most abundant in the majority of leaves? _____

8. Which layer of *mesophyll* is closest to the upper epidermis of a leaf? _____

9. In what kind of leaf would you expect to find *resin canals?* _____

10. The two cells that form and surround a stoma are known as _____

Plant Propagation

© Kingsley Stern

Materials

1. Dormant willow or poplar twigs, about 1.5 centimeters in diameter
2. Dormant willow or poplar twigs, about 0.75 centimeter in diameter
3. Flats containing potting soil
4. *Rootone* or similar IAA powder
5. Polyethylene bags ("baggies") containing vermiculite
6. *Begonia* (or similar plant) stem segments that can be rooted
7. Rose cuttings
8. Carrots
9. White potatoes
10. Sweet potatoes
11. Grafting wax/sealing compound
12. Facial tissues
13. Wide-mouthed jars (Mason or equivalent)
14. Knife
15. Petri dishes
16. Toothpicks
17. Live African violet or *Peperomia* plants
18. Enamel pans (for African violet/*Peperomia*s)
19. Roll of aluminum foil
20. Bulbs (e.g., lily)
21. Lily plant with axillary bulbils
22. Pineapple (whole)
23. Clay or plastic pots with potting soil
24. 10% liquid bleach solution
25. Ripe apples
26. Long forceps
27. Nutrient agar slants
28. Test-tube rack
29. Bunsen burner
30. Matches
31. Flat rubber strips or plastic tape (for grafts)
32. Single-edged razor blades
33. Melon-ball scoop
34. Light banks (for maintenance of propagated materials in the laboratory)
35. Vermiculite
36. Containers for plant bulbs
37. Several *Equisetum* stems
38. Raw peanuts
39. Models and charts

Some Suggested Learning Goals

1. Be able, using simple techniques, to perform propagation of new plants from stems, roots, and leaves.
2. Experience and understand the technique of *budding* (bud grafting), and know why grafting can and does occur.
3. Experience and understand the propagation on an artificial medium of a plant from a seed embryo, and know why only sterile techniques should be employed.
4. Be able, using formal English, to write up the embryo culture experiment in scientific format.

Introduction

In nature, most plant species are perpetuated through propagation by means of seeds or spores. Crops and ornamental plants also have been propagated by means of seeds or spores, harvested either from wild plants or from those grown under controlled conditions, ever since earlier peoples first developed agricultural techniques.

However, genetic factors involved in the production of seeds or spores may cause plants grown from them to vary from plant to plant. Some plants such as orchids may grow very slowly and take several years to progress from seed to the first flowers. *Vegetative propagation* methods are frequently used to reduce the problems of genetic variation as well as to speed up the propagation of certain ornamentals. Such methods include the use of *cuttings, divisions, grafting, layering,* and *tissue culture*. This exercise will introduce you to some of these techniques.

A. Techniques

1. Using a clean, sharp razor blade or knife, cut pieces of the twigs provided into 15-centimeter (6-inch) lengths. Remove all except the top one or two buds or leaves. Taking care to have the tip end pointing up, insert the bottom end of the twig lengths 2.5 to 5 centimeters (1 to 2 inches) into the vermiculite or other culture material provided. Dust hormone powder on the wetted bottoms of half of the twig lengths before inserting them, and keep the treated ones in separate containers. Water both groups of twigs each week until obvious

growth has occurred. Then dig up the twigs and compare root production between the treated group and the untreated group.

2. Cut off the top 1 to 2 centimeters (0.5 inch) of a carrot provided, and place it in a petri dish. Add enough water to cover all except the original leaf bases, and do not allow it to dry out for at least 1 month. **Important:** *Change the water twice a week.*

3. Suspend half of a sweet potato with toothpicks in a jar so that most of it projects down into the jar. Fill with water and set aside. **Important:** *Change the water twice a week.*

4. Cut a white (Irish) potato into pieces so that each piece has at least one good "eye." Plant the pieces in the pot provided. Water twice a week.

5. Place foil or wax paper across the mouth of a glass jar or pan filled with water, and secure the foil or paper with a rubber band. Gently poke small holes in the foil. Insert the petioles of African violet leaves through the holes, dusting half the petioles with hormone powder before insertion. Use separate containers for the treated leaves. Compare results in 6 weeks. Be sure to maintain water levels while waiting for results.

6. Punch out or cut out discs of *Rex Begonia* leaf blades, and place them on damp filter paper in petri dishes. The discs should be about 2 centimeters (0.8 inch) in diameter and should include parts of the larger veins. Maintain the dampness of the filter paper, and place in an area that receives a minimum of several hours of light (not direct sunlight) per day. Check weekly for the development of roots.

7. Take a lily plant with flowers in the bud stage, if available, and carefully remove all the flower buds. Set the plant aside, and watch for the development of bulbils (small bulbs) in the leaf axils. These bulbils may be removed and planted separately.

8. Cut the top off the pineapple provided, and plant it in a pot of well-drained soil. When sufficient growth has occurred (e.g., after several weeks when the leaves are about 30 centimeters [1 foot] long), flowering may be induced by placing an apple in the pot (to provide ethylene gas) for 2–3 days and covering the plant with a clear plastic bag. Be sure to water regularly.

9. Cut horsetail (*Equisetum*) stems into segments about 2.5 centimeters (1 inch) long. Each segment should contain a node. Float the segments in water for 2 weeks, and check for development of *adventitious roots* (i.e., roots developing directly from the cuts or sides of the stem).

B. Embryo Culture

This technique has been used to speed up the propagation of seeds that do not germinate readily, and it also increases the growth rate of the seedlings. To be successful, this type of propagation must be performed under aseptic conditions. Sterilize the surface of the seeds by soaking them for 10 minutes in a 10% solution of liquid bleach. Then, using a sterilized probe and razor blade, remove the embryo from the peanut, apple, or other seed provided, and plant it in a sterilized test tube containing a slant of nutrient *agar*.

Agar, which has the consistency of gelatin in both its cooled solid form and its liquid heated form, is obtained from seaweeds. It lends itself well to enriching with various nutrients and other materials. The agar in the test tubes provided is slanted because after solid agar had been sterilized and dissolved in hot water, the tubes were cooled while inclined at an angle. This provides a more extensive surface area for growth than is possible when the agar is cooled while the tubes are vertical.

To ensure a sterile surface to the mouth of a test tube after handling, insert it in a flame before replacing the cotton plug, and place the test tube in subdued (but not too shaded) light. If contamination (evidenced by the growth of fungi or bacteria on the agar) does not occur and the technique is otherwise successful, the seedling that develops may be transplanted to soil in about 4 weeks. If a transfer room equipped with ultraviolet light for sterilization is available for the process, the chances for success might be enhanced.

C. Tissue Culture

A demonstration of this process may be available. Your instructor will explain the steps to you. This technique is now employed commercially, particularly by large orchid growers.

D. Grafting

There are many different kinds of grafts used for a variety of purposes. You will perform two of them in the laboratory.

1. *Budding (bud-grafting)* (Fig. 7.1). Using a sharp razor blade, cut a T (the T may be either inverted or upright) in the side of a stout twig. The cuts should go no deeper than the cambium, which appears as a moist, white to green layer, possibly less than 3 millimeters below the surface. At the same time, carefully cut into the bark around a bud on another twig (allow about 1 centimeter [0.4 inch] or more on all sides of the bud), again going no deeper than the cambium. Carefully fold back the flaps of tissue at the T cuts, and insert the bud. Now close the T around the bud, tie it shut with the material provided, and seal the graft with grafting wax or sealer to keep it from drying out and to exclude bacteria and other disease organisms as much as possible. Any new tissues will be produced by the cambium of both parts of the graft, and both cambiums *must* be in contact for the graft to succeed. Keep the base of the twig immersed in water while waiting for the graft to "take."

2. *Whip or cleft-grafting* (Fig. 7.2). Cut a V-shaped notch in the top of a stout twig, in the region of the cambium.

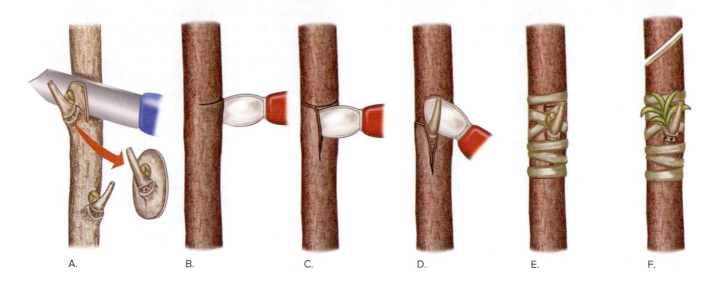

FIGURE 7.1 STAGES IN BUDDING (BUD-GRAFTING).

A.   B.   C.   D.   E.   F.

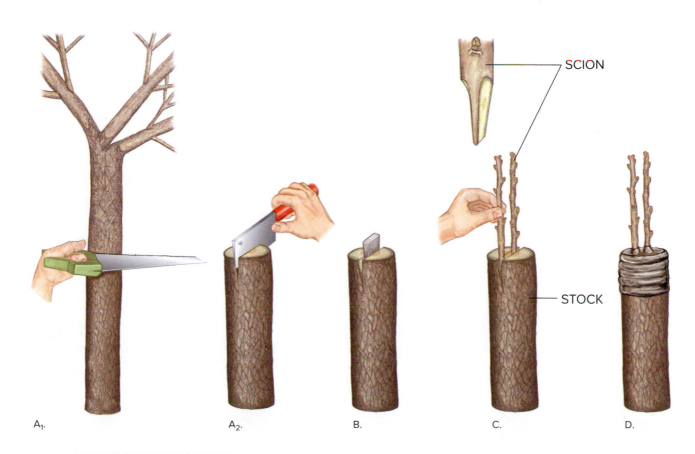

SCION

STOCK

A₁.   A₂.   B.   C.   D.

FIGURE 7.2 STAGES IN CLEFT-GRAFTING.

Now cut the base of a smaller twig to match the notch. Insert this smaller twig in the notch, and seal the graft with grafting wax to prevent tissues from drying and to reduce the chance of bacterial infection. Again, keep the base of the twig immersed in water until growth appears.

Optional Tour

When the propagation exercises have been completed, your instructor may take you on a tour of a greenhouse or arboretum. Take notes during the tour, and follow the instructor's directions carefully.

Assignment

Prepare a scientific report of the embryo culture experiment in appropriate formal English, using the headings and outline that follows. You will not be able to complete the report until after results have been obtained—usually within 2 weeks.

1. **Title.** Provide a title that describes the investigation(s) being pursued. This description should not be a mere restatement of the laboratory topic (e.g., "Plant Propagation") nor should it be so long that it exceeds more than one sentence. The title is a statment that accurately identifies the experiment.

2. **Introduction.** Explain the purpose of the experiment and what new information should be revealed from the investigation(s). An introduction may contain background information about the topic and how previous work has led to the proposed experiment. A *hypothesis* is simply a statement that an experiment is designed to test; the statement itself may or may not be true. A hypothesis should be included to suggest possible outcomes of the experiment. Typically, this hypothesis is written in a manner that can be easily tested. For instance, a hypothesis might be, "It is possible to culture plants from seed embryos on an artificial medium." If the experiment reveals that such a possibility exists, then it can be concluded (see "Conclusions," described below) that the results support the hypothesis. However, if the results contradict the hypothesis, then an appropriate statement should be made in the conclusions.

3. **Materials and methods.** *List* or describe the materials and equipment used, and tell exactly how the experiment was set up (with the aid of diagrams). If appropriate, use a *control*. A control is an experiment that is managed exactly as all others, except for treatments that are being investigated to test the hypothesis. In this respect, the control provides a check on the validity of what happens in the other experiments. In the embryo culture exercise, a control might consist of an experiment performed in exactly the same way, but without sterilizing the seed surfaces, forceps, and razor blade; *or,* the agar of the control could be prepared without nutrients.

4. **Results and discussion.** Tell exactly what happened, including anything unexpected. Include observations, drawings, and any numerical results in this section. This is an exciting part of the scientific report because it provides an opportunity to convey new information. If you have reason to believe something went wrong, discuss it, and suggest reasons for the unexpected occurrences.

5. **Conclusions.** *Be sure your conclusions are related to the hypothesis you stated in the introduction.* State whether or not the results support the hypothesis. Provide justification of your conclusion and how it contributes to current understanding of the topic.

Review Questions 7

NAME _____

LAB SECTION NO. _____

DATE _____

1. Why do plants grown from seed vary more than those propagated vegetatively? _____

   _____

2. What common methods of *vegetative propagation* are discussed in this exercise? _____

   _____

3. When making *cuttings*, which end should be inserted in the soil? _____

4. What parts of the following plants are used for propagation in this exercise? carrot: _____

   sweet potato: _____

   white potato: _____ African violet or *Peperomia:* _____

   Rex Begonia: _____ pineapple: _____

5. Why should water be changed frequently in some of the experiments performed in this exercise? _____

   _____

6. What is *agar*, and why is it used in the embryo culture experiment? _____

7. For a graft to be successful, which specific tissue in both parts of the graft must be in contact? _____

   _____

8. What useful purpose does *grafting wax* or *sealer* serve*?* _____

   _____

9. What is a *hypothesis?* _____

10. What is a *control?* _____

Laboratory Preparation
Quiz 7

NAME _____

LAB SECTION NO. _____

DATE _____

Plant Propagation

1. List three methods of *vegetative propagation*. _____

2. When making *cuttings*, what should be removed from the twigs? _____

3. What is dusted on the bottom of some of the cuttings in this exercise? _____

   _____

4. When making divisions of white (Irish) potatoes, what structure(s) must be present if the division is to sprout? _____

   _____

5. To which part of an African violet or *Peperomia* leaf is hormone applied in this exercise? _____

   _____

6. What, theoretically at least, does *embryo culture* do to seedlings? _____

7. How are seed surfaces sterilized in this exercise? _____

8. Which vegetative propagation technique referred to in this exercise is used by large commercial orchid growers?

   _____

9. In making a *bud graft*, how deeply should the T be cut? _____

10. What is a *hypothesis?* _____

Cell Components and Products

© Kingsley Stern

Materials

1. Starch solution
2. Glucose solution
3. Protein solution
4. Thin Brazil nut slices floating in water
5. Solutions of unknowns
6. Forceps
7. Large beakers (400 milliliter)
8. Small beakers (50 milliliter)
9. Petri dishes
10. Test tubes
11. Test-tube holders
12. Mortar and pestle
13. Hot-water baths
14. Tripods and insulated wire gauze for hot-water baths
15. Test-tube brushes and dishpan with soapy water
16. IKI reagent
17. Protein alkaline solution
18. Tes-tape
19. Biuret reagent
20. Sudan IV dye
21. Dilute HCl (hydrochloric acid) or vinegar solution
22. Dilute NaOH (sodium hydroxide) or ammonium hydroxide solution
23. Bunsen burner
24. Matches
25. Pots with lids
26. Trays of ice cubes
27. Facial tissues
28. Filter paper
29. 95% ethyl alcohol
30. Peppermint, pennyroyal, or other mint plants
31. Head of purple or red cabbage

Some Suggested Learning Goals

1. Know tests for the presence of starch, sugar, and protein. Be able to use the tests to determine the presence or absence of these substances in unknown solutions.
2. Know a simple test for the presence of fats or oils.
3. Know a simple technique for removing oils from leaves.
4. Understand how pH affects the color of anthocyanin.

Introduction

Apart from water, *carbohydrates, fats,* and *proteins* constitute a large proportion of a cell's makeup, but there are literally hundreds of other identifiable compounds found in cells. Those useful to humans include numerous medicinal drugs, gums, resins, spices, condiments, cellulose, and various pigments. The presence of some of these compounds may be determined by relatively simple tests. In this exercise, you will perform some of these tests and will also note the localization of some of the substances. Reagents for simple tests include (1) an iodine solution that tests for presence of starch; (2) Tes-tape, used to detect glucose; (3) Biuret reagent, which is sensitive to protein; and (4) Sudan IV, which stains fats.

Some plant compounds, such as anthocyanin (a water-soluble pigment generally found in the vacuole), are sensitive to pH. The color of this pigment can be changed by the addition of acid or base to plant tissues. A simple experiment may be performed by changing the pH of plant extracts containing anthocyanin.

A. Food Substances

1. Set up nine test tubes in three groups of three test tubes each. Add about 2.5 centimeters (1 inch) of starch solution to each test tube in the first group; add a similar quantity of protein solution to each of the three test tubes in the second group, and a similar quantity of sugar solution to each of the three test tubes in the third group. Now add 2 to 3 drops of iodine reagent to one test tube in each of the three groups. In which test tube does an inky-black color change occur? Next, repeat the process by dipping one of three 5-centimeter (2-inch) lengths of Tes-tape for no more than 5 seconds halfway in another test tube in each of the three groups. Then remove each of the three Tes-tape strips, and inspect the damp portions for obvious changes in color from yellow to green. Finally, add 30 drops of alkaline reagent to each of the remaining three test tubes, followed by 5 drops of Biuret reagent. Again, note the *most obvious* color changes (usually a clear, pale lilac). For which food substances are the various reagents specific? Test the unknown solutions, if available, as your instructor directs you.

2. Your instructor may choose to use a microscope to demonstrate oils in Brazil nuts. When a drop of *Sudan IV,* a reddish oil-soluble dye, is added to a thin sliver of Brazil nut, the small droplets of oil become clearly visible. Should your instructor direct you to do the demonstration yourself, be sure to avoid skin contact with Sudan IV and wash immediately if an accident should occur, because the dye is believed to be carcinogenic.

Thin slivers of Brazil nut floating in water may be provided. Place one sliver on a *clean* microscope slide. Your sliver may curl up if you try to remove it with forceps; if so, try to slide the tip of your microscope slide under the sliver in the water in which it is floating, and then *slowly* raise it up. Add a drop of Sudan IV. Observe with your microscope under low power. Because Sudan IV is a fat-soluble dye, you would expect it to stain any fats or oils present. Does the dye appear to be localized in small oil (fatty) droplets present in the Brazil nut sliver?

B. Aromatic Oils

Members of several plant families, notably the Mint Family (Lamiaceae) and the Laurel Family (Lauraceae), produce aromatic oils in the leaves and other parts of the plants. The oils can be extracted by relatively simple techniques. Your instructor may choose to demonstrate a technique to the class, or he/she may have you form small groups, with each group extracting oils themselves.

Place two dozen or more leaves of fresh mint (peppermint, pennyroyal, spearmint, etc.) in the bottom of a large pot or beaker that has a wire rack or other support for a small bowl. Add just enough water to cover the leaves, but don't have the water extend above the top of the rack (Fig. 8.1). Place a small empty bowl on the rack, and bring the water to a boil. As soon as the water begins to boil (do *not* allow the water to boil for any length of time), invert the pot lid or other concave cover over the pot or container, and dump up to two dozen ice cubes in the center of the inverted lid. As the oil, now in vapor form, circulates within the container, it comes in contact with the cold area on the lid, condenses, and drips into the bowl. Some water may also condense, but because the oil is lighter than water, it will float on top and can be removed by carefully pouring it off or extracting it with an eyedropper. (Mint oil can be refrigerated in a closed container and used indefinitely as a flavoring, if desired.)

C. Plant Pigments

Place a few leaves of purple cabbage into a mortar and pestle with 10 milliliters of water, and grind for 1 minute to extract the water-soluble anthocyanin pigment. Decant 3–5 milliliters of the resulting liquid into each of two 50-milliliter beakers. Note the color of these solutions, which should both be similar in color to that of the original tissue. To one beaker, add drops of dilute HCl (vinegar may be substituted if HCl is not available), and note the change to a red or reddish purple color. To the other beaker, add several drops of dilute NaOH (dilute ammonium hydroxide may be substituted if NaOH is not available), and note the change to a blue or blue-green color (Fig. 8.2). What does this tell you about the sensitivity of anthocyanin to pH? Does this help explain why the color of purple cabbage can vary, depending on the pH of the soil where the plant is grown?

FIGURE 8.1 A SIMPLE APPARATUS FOR DISTILLING MINT OIL AT HOME.

FIGURE 8.2 VARIATIONS IN COLOR OF ANTHOCYANIN EXTRACTED FROM PURPLE CABBAGE.
Courtesy of James E. Bidlack

Exercise 8

NAME _____

LAB SECTION NO. _____

DATE _____

Food Substances

Reagent	Color of Reagent	Color Produced with			Reagent Specific for
		Starch	Glucose	Protein	
Iodine (IKI)					
Tes-tape					
Biuret					

Unknowns
(Use + for present and − for absent)

Number of Unknowns	Starch	Glucose	Protein
1			
2			
3			
4			
5			
6			

Review Questions 8

NAME _____

LAB SECTION NO. _____

DATE _____

1. In this exercise, which material or substance is used to test an unknown solution for the presence of sugar? _____

   _____

2. How is the test for sugar performed? _____

   _____

3. What is the equivalent test for starch? _____

   _____

4. What is the equivalent test for protein? _____

   _____

5. What is the equivalent test for oil? _____

   _____

6. Where, specifically, is the Sudan IV dye located in the Brazil nut slivers? _____

   _____

7. What property of an oil enables it to be separated from water? _____

   _____

8. How do you think the color of purple cabbage would be affected if the plant were grown in an acid soil? _____

   _____

Laboratory Preparation
Quiz 8

NAME _____

LAB SECTION NO. _____

DATE _____

Cell Components and Products

1. Other than starch and sugar solutions, what solution is tested in this exercise? _____

2. In addition to iodine reagent and Biuret reagent, what testing material is used in the food testing part of this exercise?

   _____

3. What fat-soluble dye is used in testing Brazil nut slivers? _____

4. What color changes are produced with the tests? _____

   _____

5. When testing a solution with Tes-tape, how long should a strip be held in the solution before being removed? _____

   _____

6. When a clear, pale lilac color is produced, what reagent(s) is (are) involved? _____

7. Name a plant family whose members are known for producing aromatic oils. _____

8. What is the function of the ice cubes in the extraction of oil from leaves? _____

9. What property of oils allows them to be separated from water relatively easily? _____

10. What is the purpose of having a rack in the pot used for the extraction of oil from leaves? _____

    _____

11. What water-soluble pigment is extracted from purple cabbage when it is mixed with water and ground with a mortar

    and pestle? _____

Diffusion, Growth, and Hormones

Materials

1. India ink
2. Agar in petri dishes for crystal dye diffusion
3. Potassium permanganate crystals
4. Green or blue dye crystals
5. Slides and coverslips
6. 25% sucrose solution
7. IKI solution
8. Healthy *Elodea* plant in water
9. Ball of starch-free string
10. Scissors
11. Matches
12. Tie-on plant tags
13. Grid marker and ink pad
14. Aqueous eosin or food coloring
15. Beakers (600 milliliter)
16. IAA in lanolin
17. Lanolin
18. *Rootone* (contains auxin)
19. Toothpicks
20. Single-edged razor blades
21. Small *Coleus* plants
22. Dropper bottles with pure water, 0.1 microgram gibberellic acid solution, and 1.0 microgram gibberellic acid solution
23. 3 membrane-covered thistle tubes or similar osmometers with ring stands, clamps, and 400-milliliter beakers—for osmosis demonstration
24. Glass cylinder with stopper, ring stand with clamp, flask of ammonium hydroxide, litmus paper, and meter ruler—for diffusion demonstration
25. 100-milliliter beaker, presoaked dialysis tubing, dialyzing solution (e.g., soluble starch and glucose solution)—for dialysis demonstration
26. Transpiration demonstration
27. *Mimosa pudica* and insectivorous plants—for turgor movement demonstrations
28. Phototropism demonstration
29. Gravitropism demonstration
30. Etiolation demonstration
31. Two bell jars with holly twigs beneath, one with a ripe apple, the other without—for effect of ethylene on abscission demonstration
32. Balsam (*Impatiens*) plant
33. Potting soil and pots for planting

Some Suggested Learning Goals

1. Understand what *Brownian movement* is and why it occurs.
2. Know and understand *diffusion, osmosis, dialysis,* and *plasmolysis.*
3. Understand and be able to explain *transpiration* and *turgor movements* in leaves.
4. Understand and be able to explain how *auxin* and *gibberellin* affect stem growth; what *tropisms* are and why they occur; *etiolation;* and the effect of *ethylene* on leaf abscission.

Introduction

Water is vital to the growth, health, and survival of plants and animals alike. Plants take up most of the water they need through osmosis and release it back into the atmosphere by the process of transpiration.

Plant hormones influence or regulate various metabolic processes and are produced in minute amounts, usually in specific areas such as meristems. They then diffuse or are transported elsewhere where they influence growth and flowering. Most plant hormones produce their results in concentrations of as little as a few parts per million. Scientists have found and are still finding many uses for these plant hormones in the production of more uniform and larger crops and in a myriad of other horticultural practices.

The following experiments are designed to demonstrate the movement of water through membranes such as the plasma membrane of cells and to illustrate the release of water in vapor form from plants. Other experiments demonstrate various uses and activities of plant hormones.

Plants and Water

A. Brownian Movement

Mount a drop of diluted dye or diluted India ink on a microscope slide. Add a coverslip, and observe under high power. The particles are moving, not because there is life in them but because the energy of the molecules making up the fluid in which they are suspended is causing the molecules to bump into the visible particles. Now remove your slide, and gently heat it for a few seconds with a match. Observe again. Does temperature have anything to do with the energy of the fluid molecules?

B. Diffusion

Diffusion may be defined in terms of the tendency of molecules to distribute themselves evenly in the space available. In diffusion, molecules or ions move from a region of higher concentration to a region of lower concentration until a state of equilibrium is reached.

Place a crystal of dye and/or potassium permanganate on the surface of an agar plate. If you use more than one kind of crystal, be sure they are approximately the same size and that they are placed about 2.5 centimeters (1 inch) apart on the agar. Check the crystals at 5-minute intervals. Record your results. Your instructor will demonstrate diffusion of molecules through air in a tube. Also record those results.

C. Osmosis

Osmosis may be defined as the diffusion of a *solvent* (e.g., water) through a semipermeable membrane (Fig. 9.1). The plasma membranes of living root cells of plants are semipermeable membranes that function in osmosis.

Observe the three osmometers that have been set up as a demonstration. An osmometer is any device, such as a thistle tube with a differentially permeable membrane across its mouth, that will permit osmosis to occur. Note the levels of the colored fluids in the tubes at the start of the demonstration and at four successive 15-minute intervals after that. Record your results. The fluids involved are shown below in Figure 9.1.

D. Dialysis

The diffusion of a *solute* (something that is dissolved in a fluid) through a differentially membrane is called *dialysis*.

Observe the demonstration of dialysis, in which dialysis tubing resembling sausage casing is used. The dialysis tubing is differentially permeable and was tied off at one end. The bag thus formed was then filled with a mixture of liquid starch and glucose solution. After washing the bag carefully to remove any spilled solution, its other end was then also tied off. Following this, the bag was immersed in a beaker containing just enough tap water to cover the bag. Both the bag and the water in the beaker can be tested for starch (using an iodine solution) and sugar (using Tes-tape) toward the end of the laboratory period. Record the results.

E. Plasmolysis

The protoplasm of a cell will shrink away from the cell wall if water passes out of the cell. This phenomenon is known as *plasmolysis,* and it will occur if the concentration of water molecules inside the cell is higher than the concentration of water molecules outside the cell. Mount a leaf of *Elodea* in a drop of tap water on a microscope slide, and briefly observe the cells under high power. Now add a drop of 25% sugar solution to the side of the coverslip, and draw it under the coverslip by touching a tissue or a paper towel to the opposite side. After a minute or two, observe the cells again. What has happened?

F. Transpiration

Although some leaves with special modifications allow water to leave the plant in liquid form, most of the water that enters a plant by osmosis leaves the plant as invisible vapor through the stomata. Examine the demonstration of condensed water vapor on the sides of a jar that has been inverted over a transpiring plant.

G. Tissues Involved in Water Transport

Cut off the root system of a balsam (*Impatiens*) or similar plant, and quickly immerse the cut end in a beaker of *eosin* (a nontoxic water-soluble dye). After 15 minutes, using a sharp razor blade, carefully cut *thin* cross sections of the stem, mount the thinnest section in water on a slide, and examine under the microscope.

H. Turgor Movements

Examine the demonstration of plants that exhibit turgor movements. The sensitive plant (*Mimosa pudica*) has leaves with a swelling called a *pulvinus* at the base of each petiole; there is also a small pulvinus at the base of each leaflet. Note the rapid response when a leaf is touched. Water in the cells of half of each pulvinus runs out into intercellular spaces, causing those cells to lose *turgor* (firmness due to internal water pressure), and a movement results. As water slowly returns to the cells, the leaf and leaflets assume their original positions. Your instructor will explain the movements in any of the insectivorous plants on display.

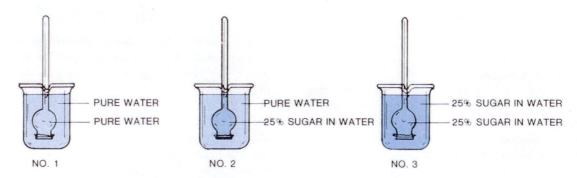

NO. 1   NO. 2   NO. 3

FIGURE 9.1 DEMONSTRATION OF OSMOSIS.

Plant Hormones

There are several known types of *plant hormones* (also called *plant-growth regulators*). These include *auxins, gibberellins, cytokinins, abscissic acid,* and *ethylene.* Some of the plant hormones produce their effects independently; others may enhance or inhibit another hormone's effects as they interact with each other. The following includes descriptions of simple experiments that can be used to demonstrate the effect(s) of hormones on plant growth.

I. Stimulation of Root Formation by Auxin

Select two similar fresh *Coleus* plant stems that measure between 1 and 2 inches in length from tip to base. With a razor, cut the stems at the base, and carefully remove any leaves within a half-inch of the cut. Dip the base of one cutting into *Rootone* (which contains auxin), and then transfer it to a pot containing soil mix prepared by the instructor. Place the other cutting directly into another pot with the same soil mix. Label the pots as "treated with auxin" or "control" in order to keep track of the plants. Keep the soil moist for 3–4 weeks, and observe any formation of roots at the end of the experiment. Record which plants produced more root growth, and explain the effect of auxin on root formation.

J. Effect of Auxin on Stems

Apply a small amount of *auxin* mixed with lanolin to one side of the stem of a *Coleus* plant, just below the growing tip. Place an identification tag on the plant, and set it where it will receive light. Place a control plant, to which lanolin without auxin has been applied, alongside it. Observe and record results after 2 or 3 weeks.

K. Effect of Gibberellin on Stems

You have been furnished with two dropper bottles of *gibberellin* and one dropper bottle of water, to each of which a *surfactant* (wetting agent) has been added. Apply several drops of 0.1 microgram solution of gibberellin to the leaves of a supplied young genetically dwarf corn plant. Apply several drops of 1.0 microgram solution of gibberellic acid to a second dwarf corn plant, and apply several drops of pure water to a third corn plant (the *control*). After 1 week, examine the plants. Does externally applied gibberellin promote stem growth in dwarf plants? Why was it necessary to add a surfactant to the solutions? Is there a difference in the effects of dilute versus more concentrated gibberellin on stem growth?

L. Growth Rates of Plants

Carefully mark different sections of a healthy plant with India ink, using the device provided (the strings are exactly 1 millimeter apart). Record your treatment, and 2 weeks later record any differences.

M. Effect of Terminal Bud Hormones on Growth

Certain hormones produced within terminal buds inhibit the development of axillary or lateral buds. Remove the terminal bud from a healthy *Coleus* plant, and place an intact similar plant alongside as a control. Compare the growth of the two plants after 2 to 3 weeks. Record your results.

N. Phototropism

Examine the plants that for a few days were kept under a box with a hole admitting light from one side. They are showing a *positive phototropic* response (bending in the direction of the light) because the cells on the dark side of the plants have elongated more than the cells facing the light. What hormone is responsible for the greater elongation of cells on the dark side? Is a positive phototropic response of any value to the plant's survival?

O. Gravitropism

A growth response of a plant organ to the stimulus of gravity is referred to as a *gravitropism.* Plant organs such as stems that grow upward are said to be *negatively gravitropic,* and roots that grow downward are said to be *positively gravitropic.* Any bending of plant organs in response to gravity is due, as in phototropic responses, to greater elongation of cells on one side of the organ than on the other. There is evidence that the same concentration of auxin that stimulates cell elongation in stems may inhibit cell elongation in roots. Examine the potted plant that had been tipped on its side a day or two earlier. Explain the bending that you see.

P. Etiolation

Seeds that germinate in the dark produce stems that elongate rapidly, pushing the shoot up through the dark soil, with the leaves remaining small, and the whole plant appearing pale and spindly. Such a plant is said to be *etiolated.* Chlorophyll does not develop until the plant is exposed to light, at which time *phytochrome* pigment plays a role in triggering normal growth. Examine the demonstration of etiolated plants (plants grown in the dark) and similar normal plants germinated at the same time and grown in the light. Record differences in lengths and sizes of the plant organs. Could there be any advantage to a plant that germinates in dim light to grow in an etiolated condition?

Q. Effect of Ethylene Gas on Abscission

Observe the two holly branches that have been set up in beakers of water, one under a bell jar with a ripe apple (the apple produces ethylene gas as it ripens), the other (*control*) under a bell jar without an apple. Record any changes you observe 1 week later.

Exercise 9

NAME _____

LAB SECTION NO. _____

DATE _____

Review Questions 9

NAME _____

LAB SECTION NO. _____

DATE _____

1. What is *Brownian movement?* _____

   How does temperature affect it? _____

2. What causes dye to diffuse away from its source on an agar plate? _____

   _____

3. If an *osmometer* contains pure water and it is immersed in a beaker containing 25% sugar solution, what will happen to

   the level of the fluid in the tube? _____

   Why? _____

   _____

4. What is *dialysis?* _____

   Why was there no starch in the beaker at the end of the dialysis demonstration? _____

   _____

5. What happens to cell vacuoles in a *plasmolyzed* cell? _____

   Why? _____

6. When *transpiration* occurs, how and where does it take place? _____

   _____

7. When *auxin* is applied to the side of a *Coleus* stem near the tip, what do you expect to happen? _____

   _____

   Why? _____

8. When ink marks exactly 1 millimeter apart are applied to the side of a young *Coleus* stem, what do you expect to happen

   over the ensuing 2 to 3 weeks? _____

9. In the experiment with eosin and balsam (*Impatiens*) stems, in which tissue would you expect to find the dye? _____

   _____

   Why? _____

10. Explain why the various plants appear the way they do in each of the following demonstrations:

Phototropism: _____

_____

Gravitropism: _____

_____

Etiolation: _____

_____

Holly leaf abscission: _____

_____

Leaf movement in the sensitive plant (*Mimosa pudica*): _____

_____

_____

Exercise 9

NAME _____

LAB SECTION NO. _____

DATE _____

Assignment

Record the results of the various demonstrations, and answer the questions below.

Diffusion of Dye Molecules

Time in Minutes	Distance in Millimeters
5	
10	
15	
20	

Osmosis

Time in Minutes	Fluid Rise, in Centimeters		
	No. 1	No. 2	No. 3
15			
30			
45			
60			

1. Could the rate of diffusion of the dye molecules have been slowed down or sped up in any way? _____

   How? _____

   _____

2. Why was there no change in the fluid levels of osmometers No. 1 and No. 3? _____

   _____

   _____

3. At the end of the dialysis experiment, why was there no starch in the beaker solution? _____

   _____

   _____

4. If you place sticks of celery or carrots in a solution of 10% salt water, should they become crisp and crunchy or limp and bendable? _____

   Why? _____

   _____

5. In which tissue of the balsam (*Impatiens*) stems is the dye located? _____

6. Make small sketches of the demonstrations and explain them on the back of this sheet.

Laboratory Preparation
Quiz 9

NAME _____

LAB SECTION NO. _____

DATE _____

Diffusion, Growth, and Hormones

1. What do we call the movement of India ink particles that is observed with the high power of the microscope? _____

   _____

2. What do we call the state reached when diffusion ceases because molecules are evenly dispersed in the space available to

   them? _____

3. What name do we give to any device that will permit us to demonstrate osmosis? _____

4. What do we call diffusion of a solute through a differentially permeable membrane? _____

5. What happens in *plasmolysis?* _____

   _____

6. What is the loss of water in vapor form from a plant called? _____

7. With what substance is *auxin* mixed to facilitate its use on stems? _____

8. As mentioned in this exercise, what does a hormone in terminal buds do to plants? _____

9. In which nontoxic dye are stems immersed so that water transport can be observed? _____

10. What fruit is used in demonstrating the effect of *ethylene gas* on *abscission* of holly leaves? _____

Photosynthesis

10

Materials

1. Phenol red solution
2. Forceps
3. Four fresh *Elodea* sprigs, two of them stored in the dark and two of them exposed to direct light for 24 hours to show differences in oxygen production
4. Two or three fresh *Elodea* sprigs to show absorption of carbon dioxide
5. *Coleus* plants with leaves that have been partially covered with black paper for several days
6. *Coleus* plants with leaves that have had normal exposure to light for several days
7. 95% ethyl alcohol
8. Petri dishes
9. IKI solution
10. Glass beakers (400 milliliter)
11. Concentrated alcoholic extract of chlorophyll for fluorescence demonstration
12. Freshly prepared green plant pigment extract in dropper bottle covered with foil and with a camel-hair brush inserted in the stopper
13. 1- by 15-centimeter paper chromatography strips
14. Rack containing numbered, stoppered test tubes, each with 1.5 centimeters of chromatography solvent
15. Hot-water baths
16. Tripods and insulated gauze for hot-water baths
17. Bunsen burners
18. Test-tube brushes and dishpan with soapy water
19. Test tubes for phenol red solution
20. Test-tube holders
21. Soda straws or glass tubing of similar diameter

Some Suggested Learning Goals

1. Know the requirements for the oxygen-generating steps of photosynthesis and how to test for this in the aquatic plant *Elodea*.
2. Understand the light requirements of photosynthesis and the meaning of fluorescence.
3. Understand and be able to perform the steps involved in an experiment to demonstrate the correlation between the presence of starch and photosynthetic activity in a *Coleus* leaf. Also be able to write up the experiment in appropriate fashion, according to the instructions given for writing up a scientific report in Exercise 7.
4. Know a simple technique for paper chromatography and the sequence in which plant pigments separate on chromatography paper.
5. Understand the experiments designed to demonstrate the role of carbon dioxide in photosynthesis and fluorescence in chlorophyll.

Introduction

Photosynthesis, the most important chemical process on earth, takes place in the presence of light, with the aid of chlorophyll pigments. During the process, carbon dioxide is combined with water, forming glucose, a simple sugar, and oxygen is produced as a by-product. The oxygen is essential to the respiration of nearly all living organisms.

Photosynthesis takes place in two successive series of steps. In the first series, called the *light-dependent reactions*, some of the light energy absorbed by chlorophyll molecules is converted to chemical energy. During the process, water molecules are split, producing hydrogen ions and electrons, with oxygen gas being released, and energy-storing ATP (adenosine triphosphate) molecules are produced. The electrons from the split water molecules are involved in the production of NADPH (nicotinamide adenine dinucelotide phosphate, reduced form), which carries hydrogen used in the second series of photosynthetic reactions known as the *light-independent reactions*.

The light-independent reactions may develop in different ways; in the most widespread type, carbon dioxide from the air is combined with a 5-carbon sugar (RuBP—ribulose bisphophate), and the combined molecules are then converted to 6-carbon sugars such as glucose ($C_6H_{12}O_6$). Energy for these reactions is furnished by ATP and NADPH produced during the light-dependent reactions.

Much of the energy captured from the light may be stored in the glucose molecules. When a period of photosynthetic activity produces more glucose than can be used, it is converted to starch and stored in chloroplasts. The starch may later be converted back to sugar and transported to various parts of the plant. The presence of starch in a leaf is considered evidence that photosynthesis has taken place.

A. Oxygen Production in Photosynthesis

Light is required for the oxygen-generating process of photosynthesis. This can be demonstrated by observing oxygen bubbles produced by the aquatic plant *Elodea*. Immerse two sprigs of *Elodea* in a beaker of water, and store in the dark. Immerse two other sprigs of *Elodea* in another beaker of water exposed to direct light. Observe for the presence of bubbles on leaves of each plant after 24 hours (Fig. 10.1). What gas might you find in these bubbles? Describe how these bubbles may be used to explain the products of the light-dependent reactions of photosynthesis.

B. Chlorophyll and Fluorescence

Pour about 5 milliliters of concentrated alcoholic extract of leaf pigment into a test tube, and shine an intense incandescent light through it, or simply hold it up to the light (transmitted light). What color is it? Now hold the test tube against a dark background (reflected light). What color is the extract? The difference in appearance is due to a phenomenon known as *fluorescence*. Fluorescence occurs when substances reradiate light of wavelengths other than those absorbed (Figs. 10.2A and B).

C. Relationship of Photosynthesis to Chlorophyll

Glucose, the primary product of photosynthesis, usually is stored in the form of starch in various parts of the plant. Most exposed parts of plants are coated with a thin film of protective wax, and the healthy cells just beneath have intact plasma membranes. Because of this, iodine reagent applied directly to a healthy leaf normally does not indicate if starch is present. However, if the leaf is killed and the wax and chlorophyll are removed, the reagent will quickly indicate if starch is present. Neither the wax nor chlorophylls are water-soluble, but they are soluble in other solvents such as alcohol. If the leaf is killed and the wax and chlorophyll are removed, the reagent should quickly indicate if starch is present.

Obtain a fresh *Coleus* leaf from a plant that has been exposed to normal light for several days. Carefully draw the outline of the leaf and also outline the colored parts.

FIGURE 10.1   OXYGEN BUBBLES PRODUCED BY THE AQUATIC PLANT, *ELODEA,* UPON ILLUMINATION BY DIRECT LIGHT.
Courtesy of James E. Bidlack

FIGURE 10.2A CHLOROPHYLL EXTRACT PRIOR TO ILLUMINATION. *Courtesy of James E. Bidlack*

FIGURE 10.2B CHLOROPHYLL EXTRACT SHOWING FLUORESCENCE UPON ILLUMINATION BY INTENSE LIGHT. *Courtesy of James E. Bidlack*

Immerse the leaf in boiling water for about 10–15 seconds to break down the plasma membranes of the cells, and transfer it to a beaker of 95% alcohol that has been brought to a boil in a hot-water bath. **(Note: Alcohol boils at a lower temperature than water and is flammable. Do not heat alcohol directly with an open flame, and also keep any open flames away from alcohol fumes!)** When most or all of the alcohol has been extracted (indicated by the leaf turning whitish), remove the leaf, and rinse in water (alcohol tends to make the leaf brittle; the water will soften it somewhat). Then place the leaf in a petri dish, and add enough iodine reagent to cover the surface. Now make another drawing of the leaf on the same page as the original drawing. Describe and record the results. Is there a correlation between the distribution of the chlorophyll and the distribution of the starch? Why?

D. Relationship of Photosynthesis to Light

Obtain a fresh *Coleus* leaf that has been partially covered with black paper for a few days. Remove the pigments in the same manner you did in the preceding paragraph. Test for starch. Is there a correlation between light and the presence of starch in a leaf? Why?

E. Relationship of Photosynthesis to Carbon Dioxide

Fill a test tube with a dilute solution of phenol red. Phenol red is an indicator dye that is red when the pH is 7 or higher (alkaline) and pale yellow when the pH is less than 7 (acid).

With a soda straw or glass tube, gently blow into the test tube until the phenol red solution changes color. The carbon dioxide in the air breathed from your lungs partially dissolves in water, forming carbonic acid, which causes a color change in the phenol red. Now empty half of the phenol red solution into a second test tube, and add a sprig of *Elodea* to one of the test tubes. Place both test tubes in bright light for 10–20 minutes. If no color change is noted, gently agitate both tubes for a few seconds, and set them aside once more. Why is there eventually a color change in the test tube with the *Elodea* but not in the other tube?

F. Paper Chromatography

To varying degrees, the pigments present in plant parts are soluble in certain solvents. A technique called *paper chromatography* can be used to separate these pigments on the basis of different molecular solubility and attraction of molecules to paper fibers. In this experiment, a band of dissolved pigment is placed toward one end of a piece of chromatography paper. The paper is then dipped in chromatography solvent, with the pigment just above, but not touching the solvent, so as to draw the solution up the paper. As the, solvent moves up the paper, the pigments separate out in bands that can be identified.

Cut a strip of filter paper about 15 to 20 centimeters (6 to 8 inches) long and about 1 centimeter (less than 0.5 inch) wide—or use the paper strips provided. With a camel-hair brush, apply a strip of pigment 2 to 3 millimeters (about 0.1 inch) wide across the paper strip, *about 2.5 centimeters (1 inch) from the bottom*. Repeat this step at least five times, being very careful to apply the pigment in exactly the same place each time.

If test tubes with solvent have not previously been set up for you, pour about 1 centimeter (less than 0.5 inch) of solvent (usually a mixture of petroleum ether and acetone—try not to breathe the toxic fumes) into a test tube. Dip the tip of the filter paper into the solvent, taking care to keep the pigment above the top of the solvent. Place the test tube in a rack, and stopper it. When the solvent is within 0.5 centimeter (1/5 inch) of the top of the paper strip, remove the paper and allow it to dry. **(Note: The solvents used in paper chromatography are both poisonous and flammable. Handle with care!)**

The pigments separate as follows: Closest to the original strip is a yellow-green band of *chlorophyll b;* next is a blue-green band of *chlorophyll a,* followed by *xanthophylls* (pale yellow); at the top, the golden yellow band consists of *carotenes*. If n-propanol is used instead of acetone in the chromatography solvent, the pale yellow *xanthophylls* will separate out closest to the original strip, followed by *chlorophyll b*, *chlorophyll a*, and *carotenes*.

Note: If solvent-extracted plant pigment is not available, paper chromatography can be demonstrated with soluble black ink.

Assignments

1. Record the results of the oxygen production (Part A), fluorescence (Part B), and carbon dioxide (Part E) experiments on a separate page.
2. Staple your chromatogram (Part F) to the bottom of the page on which your results are recorded. Label the bands of pigments.
3. Following the instructions for writing a scientific report given at the end of Exercise 7, prepare a formal report on the correlation between photosynthetic activity and the presence of starch in a *Coleus* leaf (Parts C and D).

Exercise 10

NAME _____

LAB SECTION NO. _____

DATE _____

Review Questions 10

NAME _____

LAB SECTION NO. _____

DATE _____

1. In which treatment, light or dark, did the aquatic plant *Elodea* produce oxygen? _____

2. What is *fluorescence?* _____

3. In what form is glucose usually stored in a leaf? _____

4. Why will iodine applied to a healthy leaf usually not reveal if starch is present? _____

5. What substance is used to extract chlorophylls? _____

6. Which has a lower boiling point: alcohol or water? _____

7. In the *Coleus* leaf experiment, what visible indication is there of all the chlorophyll having been extracted? _____

   _____

8. Why does an iodine test reveal no starch in a small central part of the blade next to the petiole in a variegated *Coleus*

   leaf? _____

9. In paper chromatography, which chlorophyll pigment—chlorophyll *a* or chlorophyll *b*—separates out closest to the top of

   the paper strip? _____

   _____

10. What happens to the carbon dioxide that is bubbled into the phenol red solution? _____

    _____

11. Assume that the *Elodea* sprig in the phenol red solution carries on photosynthesis. Does the oxygen given off have

    anything to do with the color change? _____

    _____

Laboratory Preparation
Quiz 10

NAME _____

LAB SECTION NO. _____

DATE _____

Photosynthesis

1. What gas, produced as a by-product of photosynthesis, is essential to nearly all living organisms? _____

_____

2. What term is given to substances that reradiate light of wavelengths different from those absorbed? _____

_____

3. In what form is the primary product of photosynthesis stored in a plant? _____

4. Why will applying iodine reagent directly to a healthy leaf probably not reveal the presence of starch, even when you

know some cells contain it? _____

_____

5. What types of substances need to be removed from a leaf so that it can be tested for the presence of starch? _____

_____

6. Why should alcohol be brought to a boil only in a hot-water bath? _____

_____

7. What serves as evidence that the chlorophyll has been removed from a *Coleus* leaf? _____

_____

8. What do we call the technique by which pigments are separated on a strip of paper? _____

_____

9. During the paper chromatography, which of the pigments that separate out on the paper are golden yellow? _____

_____

10. What color is phenol red in an acid solution? _____

11. How does the *Elodea* sprig bring about a color change in the phenol red solution? _____

_____

Water in Plants; Respiration; Digestion

11

Materials

1. Freshly cut carnations
2. Bottles of assorted food coloring
3. Potometer
4. Small, fresh, woody twig with broad leaves
5. Small electric fan
6. Barley seedlings under a glass jar
7. One fresh potato or several bean sprouts
8. Small quantity of viable corn kernels
9. 10% liquid bleach
10. Two thermometers in stoppers or cotton plugs
11. Two insulated bottles
12. Two wide-mouthed bottles with stoppers
13. Glass tubing
14. Rubber tubing
15. 5% formalin solution
16. Suspension of yeast in 10% sucrose solution
17. One apple
18. Baryta water
19. Hydrogen peroxide
20. Starch powder
21. Tetrazolium reagent
22. Syracuse watch glasses
23. Diastase
24. Large beakers (400 milliliter)
25. Bunsen burners
26. Tripods
27. Metal gauze
28. Matches

Some Suggested Learning Goals

1. Understand how water passes through a living plant, the role of transpiration in the movement of water, and the nature of guttation.
2. Know the essence of cellular respiration and the by-products of the process.
3. Know the effects of digestive enzymes on starch.

Introduction

If there is not enough time or appropriate equipment for all students to work through the experiments discussed in the following material, parts of this exercise may be set up as demonstrations. These show the movement of water through a plant, various aspects of cellular respiration, and the effect of digestive enzymes on starch.

A. Water Flow and Transpiration

A fun experiment that shows the process of water movement in a colorful way can be easily performed. If carnations are available, simply immerse the stem of each flower into a container of concentrated food coloring. The water, along with dye, will be absorbed by the stem of the flower and transported to the perimeter of the petals. After a few hours, the flowers will be delightfully colored for admiration and decorating purposes (Fig. 11.1). While this experiment provides a colorful way of demonstrating water movement, the following experiment allows for relative measurements of how fast the water moves through a plant.

FIGURE 11.1 CARNATION FLOWERS AFTER BEING IMMERSED IN FOOD COLORING FOR SEVERAL HOURS.
Courtesy of Joy Lauffenburger

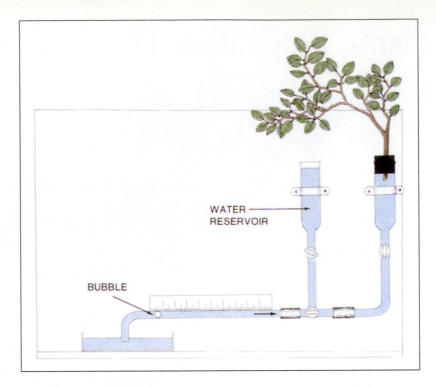

FIGURE 11.2 A POTOMETER.

Cut the base of a healthy, small branch of a broad-leaved tree under water, and then quickly insert the base into a *potometer*. A potometer (Fig. 11.2) is a device that helps us observe the rate at which water moves into a stem and *transpires* from the leaves (i.e., moves out of the leaves in vapor form). Be sure there are no air bubbles in the tubing or other glassware. When the potometer and branch have been completely set up, introduce an air bubble at the base of the tubing by lifting it out of the water reservoir for a few moments until a bubble begins to move up the tubing, then reinsert the tubing base in the water reservoir. Record the length of time it takes for the bubble to move the length of the ruler. Now introduce a second bubble, and place an electric fan so that it blows air rapidly through the leaves. Record the results in both instances.

B. Guttation

Observe the demonstration of young barley seedlings that have been grown under a glass jar to ensure the maintenance of high humidity around the plants. Note the droplets of *guttation* water at the tips of the leaves. The water has exuded from *hydathodes,* which are composed of special cells at the tips of veins where water accumulates as a result of root pressures and is forced as a liquid out of the leaf.

C. Respiration

1. Disinfect a small quantity of *viable* (capable of germinating) corn kernels by agitating them for about 30 seconds in a 10% solution of liquid bleach. Then wash the bleach off the kernels in pure water, and germinate them under sterile conditions for about 24 hours. Next, place them in a sterilized, insulated bottle (e.g., a thermos flask), and insert a thermometer that has been placed in a stopper or cotton plug. At the same time, prepare another equal quantity of corn kernels in similar fashion, but kill the kernels by boiling them for 10 minutes, and also add a disinfectant (e.g., 5% formalin) to prevent the growth of bacteria. Record the temperature in each bottle every 2 hours for 16 hours.

2. Set up two stoppered, wide-mouthed bottles as shown in Figure 11.3. Half-fill one bottle with a suspension of yeast in a 10% sugar solution and the other bottle with baryta water (a saturated solution of barium hydroxide in water). The open tube in the second bottle allows air to escape. Any carbon dioxide produced during respiration in the yeast cells will bubble through the baryta water and produce a precipitate as the carbon dioxide reacts with the barium hydroxide in the baryta water.

3. During aerobic respiration in cells, a sugar molecule is broken down and converted to pyruvic acid in the initial glycolysis phase. Then the pyruvic acid is dismantled as it runs through the citric acid cycle. While this process is taking place, hydrogen atoms are removed and passed along an electron transport chain, eventually combining with oxygen from the air, with water being formed as a result. Some hydrogen atoms, however, bypass the electron transport chain

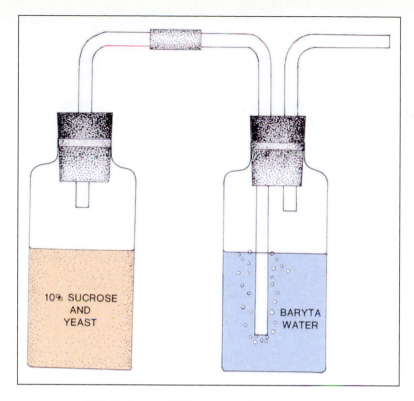

FIGURE 11.3 APPARATUS FOR MEASURING RESPIRATION IN YEAST.

and combine directly with molecular oxygen, forming *hydrogen peroxide* as follows:

$$2H \text{ (from } FADH_2) + O_2 \rightarrow H_2O_2 \text{ (hydrogen peroxide)}$$

Hydrogen peroxide *denatures* enzymes (i.e., changes the enzymes so they can't function). The hydrogen peroxide would become a major problem in cells if it were not destroyed as soon as it forms, producing water and oxygen gas in the process:

$$2H_2O_2 \xrightarrow{\text{catalase}} 2H_2O + O_2$$

Place a thin slice of apple on a clean slide, and add a drop of hydrogen peroxide to the slice. Repeat, using a drop of yeast suspension. Record your results.

D. Tetrazolium Test for Respiration

A tetrazolium reagent can be used in a quick test to determine if tissues are actively respiring. This test relies on the transfer of electrons during aerobic respiration that can be sequestered by a solution of 2,3,5-triphenyltetrazolium chloride. If tissues are actively respiring, the tetrazolium will change from colorless to pink. To perform this test, obtain some fresh potato or bean sprout tissue and add a few drops of the tetrazolium reagent. As a control, you may want to add drops of tetrazolium to another set of plant tissues that have been boiled in water for 10 minutes. The reagent takes about 15–20 minutes to begin changing color if tissues are actively respiring. In which samples (fresh tissue or boiled tissue) did you notice a color change?

E. Digestion

With the aid of a needle, stir a *small* quantity of the starch provided in a drop of water on a microscope slide, and examine with your microscope. Draw several of the starch grains you see. At the same time in a Syracuse watch glass, mix a second *small* quantity of starch in water, and add 2–3 drops of *diastase* (a starch-digesting enzyme). Place the watch glass in a warm area (preferably a low temperature oven, if available), and toward the end of the laboratory period, place a drop of the starch/diastase mixture on a clean slide. Examine with a microscope, and draw several of the partially digested starch grains.

Review Questions 11

NAME _____

LAB SECTION NO. _____

DATE _____

1. What is *transpiration?* _____

2. What is a *potometer?* _____

3. How fast did the bubble in the potometer move (a) without the fan? _____

   (b) with the fan? _____

4. What is *guttation?* _____

5. What name is given the special cells through which guttation takes place? _____

6. Record the temperature in each of the insulated bottles containing the corn kernels:

Time	Live Kernels	Dead Kernels
start:	_____	_____
2 hrs:	_____	_____
4 hrs:	_____	_____
6 hrs:	_____	_____
8 hrs:	_____	_____
10 hrs:	_____	_____
12 hrs:	_____	_____
14 hrs:	_____	_____
16 hrs:	_____	_____

7. If bacteria were allowed to grow on the dead corn kernels, how might the results differ from those obtained? _____

   _____

   Why? _____

8. What causes the white precipitate in the baryta water? _____

   _____

9. What is the function of the sugar in the yeast suspension? _____

   _____

10. What happens when hydrogen peroxide is added to fresh tissue or yeast cells? _____

    Which enzyme is involved in the reaction? _____

11. What was the purpose of including boiled tissue for the tetrazolium test for respiration? _____

    _____

12. Is there a visible change to starch grains when the enzyme diastase is added to them? What happens? _____

    _____

Laboratory Preparation
Quiz 11

NAME _____

LAB SECTION NO. _____

DATE _____

Water in Plants; Respiration; Digestion

1. What process or processes are being demonstrated with the aid of a *potometer* in this exercise? _____

   _____

2. For what purpose is an electric fan used in the first part of the exercise? _____

   _____

3. What may take place through *hydathodes* under certain environmental conditions? _____

   _____

4. In the experiment involving corn kernels, what is the purpose of adding a formalin solution to the dead seeds? _____

   _____

5. In what type of solution is the yeast suspension prepared? _____

6. What happens when carbon dioxide is bubbled through baryta water? _____

   _____

7. In this exercise, what is the source of the carbon dioxide that is bubbled through baryta water? _____

   _____

8. In living cells, which enzyme breaks up hydrogen peroxide molecules as they form? _____

9. When a drop of hydrogen peroxide is added to living cells, what gas is produced? _____

10. Would you expect the tetrazolium reagent to change color when applied to boiled tissue? _____

    _____

11. What is the function of *diastase?* _____

Meiosis and Alternation of Generations

12

Materials

1. Modeling clay (two colors)
2. Two colors and lengths of pipe cleaners
3. Sheets of brown paper
4. Models and charts of phases of meiosis

Some Suggested Learning Goals

1. Know and understand precisely what happens in each phase of *meiosis,* and be able to demonstrate the phases with modeling-clay or pipe-cleaner chromosomes.
2. Understand where gametophyte and sporophyte phases of the life cycles of plants begin and end. Also know which tissues and structures are *haploid* and which tissues and structures are *diploid.*

Before studying the events of meiosis, it is *strongly recommended* that you review and thoroughly understand *mitosis,* which was discussed in Exercise 3.

Introduction

At one stage in the normal life cycle of every organism undergoing sexual reproduction, from the simplest one-celled alga to the most complex flowering plant, certain cells undergo *meiosis.* It is a process involving two consecutive nuclear divisions, through which the original cell becomes four cells, each with half the original number of chromosomes. During the early stages of meiosis, at least some of the chromosomes may exchange parts through *crossing-over* with their *homologues* (pairs of structurally similar chromosomes, each member of a pair having been contributed by a different parent). This usually results in offspring inheriting characteristics from both parents. Meiosis is at the heart of *alternation of generations* exhibited in the life cycles of plants where the *gametophyte (n) generation* alternates with the *sporophyte (2n) generation.* In some of the simpler organisms, the sporophyte generation may consist of a single cell, the *zygote,* but in others, each phase may involve many cells and different structures.

Meiosis as Part of the Cell Cycle

Keep in mind that interphase is the longest phase of a complete cell cycle. As part of a cell cycle, meiosis follows interphase, and the individual stages of meiosis are explored in this exercise.

The following diagrams depict various steps in the process of meiosis. Although not necessarily proceeding at a uniform pace, the process is continuous from the moment the interphase cell begins to change until the four daughter cells or nuclei are in interphase themselves. In these diagrams, the original interphase nucleus has four chromosomes. One set of chromosomes (red) came from one parent, and the other set (blue) came from the other parent.

Prophase I is the longest part of meiosis. Here the chromosomes coil tightly, becoming much shorter and thicker. It becomes apparent that each chromosome consists of two *sister chromatids* held together at a constricted area by an otherwise inconspicuous *centromere.* The nuclear envelope gradually disappears, and pairing and crossing-over take place.

PROPHASE I

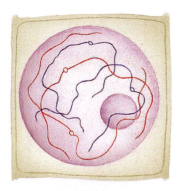

EARLY

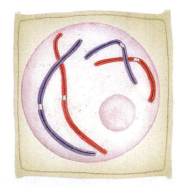

MID

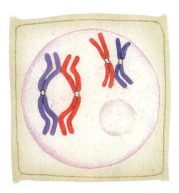

LATE

METAPHASE I

In metaphase I, the chromosomes line up in pairs along the centrally located, platelike but invisible *equator* of the cell.

ANAPHASE I

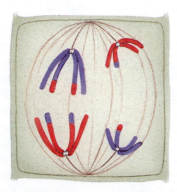

In anaphase I, one homologous chromosome from each pair migrates to a *pole* (one of two invisible points toward each end of the cell). Unlike anaphase of mitosis, *there is no separation of sister chromatids at the centromeres*.

TELOPHASE I
PROPHASE II

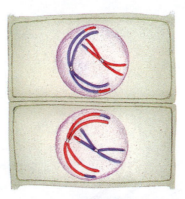

In telophase I, the chromosomes at the poles begin to become organized into two new nuclei, but because they almost immediately go into prophase II, nuclear envelopes do not normally form; visually, the effect is that of telophase activity being omitted.

METAPHASE II

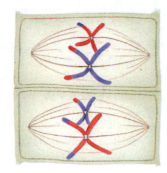

In metaphase II, chromosomes of each group become aligned along their respective invisible equators.

ANAPHASE II

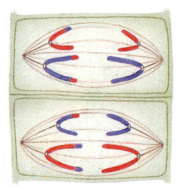

In anaphase II, the *sister chromatids* (paired strands of each chromosome) now separate longitudinally at their centromeres and migrate to opposite poles.

TELOPHASE II

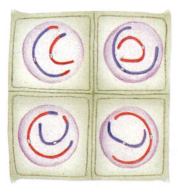

In telophase II, a nuclear envelope becomes organized around each group of chromatids that, despite consisting of single strands, are now called chromosomes again. At the same time, the chromatids uncoil, becoming longer and thinner and eventually become too diffuse to be distinguished from one another. Nucleoli appear, and cell walls usually form so that at the completion of meiosis, there are four daughter cells, each with half the original number of chromosomes.

Assignment

Using the large sheets of paper provided, draw boxes, representing cells. Make modeling-clay or pipe-cleaner models of chromosomes for each cell, and place them in their proper positions. Using the modeling clay or pipe cleaners provided, make at least four replicated chromosomes by twisting two lengths of clay or pipe cleaners around each other, to represent how *sister chromatids* are attached at interphase. The point at which the sister chromatids attach can be viewed as the *centromere* (dense constricted part of a chromosome to which a spindle fiber is attached). Use one color for chromosomes from one parent and another color for the chromosomes from the other parent. These sets of duplicated chromosomes that are the same length can be used to represent *homologous chromosomes* (chromosomes that are identical in structure and number of genes, but from different parents). During the first division of *meiosis,* the homologous chromosomes separate, resulting in a reduction in chromosome number. During the second division for each of the two cells, sister chromatids separate and migrate to opposite poles, resulting in four new cells, each of which has half the number and an unduplicated set of chromosomes within their nuclei, compared to the original cell.

To understand the stages of meiosis, refer to the figures showing individual phases, and use modeling clay or pipe cleaners to represent the chromosomes. Note that *crossing over* occurs during prophase I and *independent assortment of chromosomes* occurs during metaphase I. After the first division of meiosis (meiosis I) occurs, each of the two cells then undergoes the second division of meiosis (meiosis II). Once meiosis is completed, the chromosomes of each of the four new cells should be different from each other.

When you have completed the assignment, be sure to separate the clay or pipe cleaners into the original pure colors and store them as instructed.

Alternation of Generations

Alternation of Generations is seen throughout the Plant Kingdom, although it is more obvious in plants such as mosses, ferns, and other plants that do not produce seeds than it is in flowering and cone-bearing plants. The word *Generation* can be misleading because it is not used in the traditional sense of successive groups of offspring. Instead, *Alternation of Generations* refers to the alternation in the life cycles of plants between a *phase* in which all the cells have only one set of chromosomes (*haploid,* or *gametophyte phase*) with a *phase* in which all the cells have two sets of chromosomes (*diploid,* or *sporophyte phase*). *Gametes* (reproductive sex cells) are produced during the gametophyte phase; asexual *spores* are produced during the sporophyte phase.

Below is a diagram of a model on which any plant life cycle can be built. If you keep the following clearly in mind, you should have little trouble understanding the life cycles in the exercises to follow:

1. The first cell of any *gametophyte generation* is normally a SPORE (MEIOSPORE).
2. Any cell of a gametophyte generation is HAPLOID.
3. The first cell of any *sporophyte generation* is normally a ZYGOTE.
4. Any cell of a sporophyte generation is DIPLOID.
5. The change from a sporophyte to a gametophyte generation occurs as a result of MEIOSIS.
6. The change from a gametophyte to a sporophyte generation occurs as a result of SYNGAMY (fusion of gametes or sex cells), which is also called FERTILIZATION.

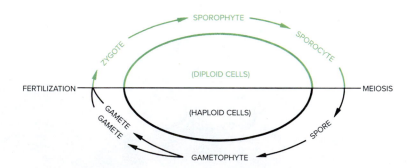

Although animals also have cells that undergo meiosis, the cells that are produced as a result of meiosis usually become gametes directly without going through an equivalent "gametophyte" phase; the terms *gametophyte* and *sporophyte* are not normally applied to animals.

Review Questions 12

NAME _____

LAB SECTION NO. _____

DATE _____

1. In which phase of meiosis does *crossing-over* occur? _____

2. In which phase of meiosis do the *sister chromatids* of the chromosomes separate completely from each other? _____
_____

3. In which phase of meiosis does a *nuclear envelope* become organized around each group of chromatids? _____
_____

4. What are chromosomal *homologues (homologous chromosomes)*? _____
_____

5. What is a cell *equator?* _____

6. Which phase of meiosis takes the longest to complete? _____

7. What name is given to the *haploid* phase of a sexual life cycle? _____
_____

8. The change from a *gametophyte* to a *sporophyte* generation occurs as a result of _____

9. What is the first cell of a sporophyte generation called? _____

10. How many sets of chromosomes are present in each cell of a sporophyte generation? _____

Laboratory Preparation
Quiz 12

NAME _____

LAB SECTION NO. _____

DATE _____

Meiosis and Alternation of Generations

1. How many divisions of the nucleus or cell occur during meiosis? _____

2. What term is applied to the exchange of parts that may occur between pairs of chromosomes in the early stages of

   meiosis? _____

3. When one cell undergoes meiosis, how many cells are there at the completion of the process? _____

   _____

4. What happens during metaphase I of meiosis? _____

   _____

5. At which stage of meiosis do the *sister chromatids* of each chromosome separate and migrate to opposite poles?

   _____

6. What is a *centromere?* _____

7. A *spore* is normally the first cell of which generation? _____

8. *Fertilization* occurs when which cells unite? _____

9. How many sets of chromosomes does a *sporocyte* have? _____

10. To which generation does a *zygote* belong? _____

Domain and Kingdom Survey

13

Materials

A display of living or preserved representatives of as many of the members of the phyla of Domains Archaea and Bacteria and Kingdoms Protista, Fungi (Mycota), and Plantae as possible.

Some Suggested Learning Goals

This exercise is an introduction to various kinds of organisms you will be studying in the exercises that follow. With the aid of your text, be able to assign the organisms on display to their correct domains, kingdoms, and phyla. *Do not be intimidated by the scientific names;* you will *not* be held responsible for them unless your instructor indicates otherwise.

Introduction

The diversity of form among living organisms is virtually infinite. Sizes range from single-celled organisms barely visible under light microscopes to complex giants like the coastal redwoods of California, which weigh many tons and form billions of cells during their growth to maturity. In order to identify, classify, and name the approximately 400,000 kinds of organisms traditionally regarded as plants, botanists utilize a hierarchy of categories, along with the Binomial System of Nomenclature, first introduced about 250 years ago by the Swedish botanist Carolus Linnaeus. The hierarchy presently recognizes eight major categories of classification (in descending order, *domain, kingdom, division* or *phylum,*[1] *class, order, family, genus, species*), with several "in-between" categories such as *subclass, subfamily,* and so on. In recent years, all living organisms have been categorized in six kingdoms, based on modes of nutrition and structure, with organisms traditionally regarded as plants occurring in five of the kingdoms. In this exercise, representatives of a number of major phyla are on display. Assign the various organisms to their correct phyla, and make notes concerning evident differences between them.

---

[1] Until 1993, botanists were required by international rules to use the term *division* as a category equivalent to phylum, which has traditionally been used for animals. An International Botanical Congress approved the use of phylum for plants and fungi in 1993.

Classification of Organisms in Domains and Kingdoms

Domain Archaea
    Phylum Archaebacteria (methane, salt, and sulfolobus bacteria)
Domain Bacteria
    Phylum Eubacteria
        Class Eubacteriae (unpigmented, purple, and green sulfur bacteria)
        Class Cyanobacteriae (cyanobacteria)
        Class Chloroxybacteriae (chloroxybacteria)
Kingdom Protista
    Phylum Chlorophyta (green algae)
    Phylum Chromophyta (yellow-green, golden brown, and brown algae)
    Phylum Rhodophyta (red algae)
    Phylum Euglenophyta (euglenoids)
    Phylum Dinophyta (dinoflagellates)
    Phylum Cryptophyta (cryptomonads)
    Phylum Prymnesiophyta (haptophytes)
    Phylum Charophyta (stoneworts)
    Phylum Myxomycota (plasmodial slime molds)
    Phylum Dictyosteliomycota (cellular slime molds)
    Phylum Oomycota (water molds)
    [Phylum Protozoa—protozoans]
    [Phylum Porifera—sponges]
Kingdom Fungi (Mycota)
    Phylum Chytridiomycota
    Phylum Zygomycota (coenocytic fungi)
    Phylum Ascomycota (sac fungi)
    Phylum Basidiomycota (club fungi)
    Phylum Deuteromycota (imperfect fungi)
    [Lichens]
Kingdom Plantae
    Phylum Hepaticophyta (liverworts)
    Phylum Anthocerotophyta (hornworts)
    Phylum Bryophyta (mosses)
    Phylum Psilotophyta (whisk ferns)
    Phylum Lycophyta (club mosses)
    Phylum Equisetophyta (horsetails)
    Phylum Polypodiophyta (ferns)
    Phylum Pinophyta (conifers)
    Phylum Ginkgophyta (*Ginkgo*)

Phylum Cycadophyta (cycads)
Phylum Gnetophyta (Gnetum, Ephedra, Welwitschia)
Phylum Magnoliophyta (flowering plants)
    Class Magnoliopsida (dicots)
    Class Liliopsida (monocots)

Kingdom Animalia (multicellular animals)

Viruses have neither cellular structure nor most of the other attributes of living organisms and do not fit into this classification. Lichens, each of which is a combination of an alga and a fungus in close association, are traditionally discussed under Phylum Ascomycota of Kingdom Fungi.

Domains (Kingdoms) Archaea and Bacteria; Kingdom Protista

14

Materials

1. Plates of gram-positive and gram-negative bacteria
2. Bacterial plates showing a variety of colonies
3. Bunsen burners
4. Live and/or preserved *Anabaena* or *Nostoc* colonies
5. Live cultures of *Ulothrix, Spirogyra, Oedogonium, Volvox, Scenedesmus, Euglena,* or *Phacus*
6. Slides of stained and preserved *Ulothrix, Spirogyra,* and *Oedogonium*
7. Diatomaceous earth
8. Herbarium specimens of seaweeds such as *Gelidium, Porphyra, Gigartina, Ulva, Codium, Postelsia, Laminaria, Costaria, Nereocystis,* or *Desmarestia*
9. Live or preserved dinoflagellates
10. Loaf of sliced white bread that contains no preservatives
11. Small petri dishes
12. Pond water
13. Dropper bottles of gentian (crystal) violet
14. Dropper bottles of safranin O dye
15. Dropper bottles of 95% ethyl alcohol
16. Live slime mold plasmodia
17. Nonexpendable and expendable slime mold sporangia
18. Preserved, dried, or living specimens of slime molds

Some Suggested Learning Goals

1. Know how to distinguish *gram-positive* bacteria from *gram-negative* bacteria, and all bacterial cells from those of Kingdom Protista.
2. Understand distinctions between *heterocysts* and *akinetes.*
3. Learn the differences among *Ulothrix, Spirogyra,* and *Oedogonium* with respect to reproduction and chloroplasts.
4. Understand how a *diatom* is constructed and how it moves.
5. Know the parts and structure of a *dinoflagellate* and of a *kelp* or other large seaweed.
6. Know how a slime mold *plasmodium* moves and the structure of slime mold sporangia.

Note: *Slime molds* (myxomycetes) are believed to be members of Kingdom Protista because they have several features commonly found in other members of this kingdom, including reproductive cells with flagella, that are not generally found in members of Kingdom Fungi. Nevertheless, because they also have certain fungus-like features, they have in the past been treated as fungi and have traditionally been studied along with members of Kingdom Fungi, discussed in the next exercise (Exercise 15). Your instructor may or may not choose to defer examining these organisms until you study Kingdom Fungi.

Introduction

Domains (Kingdoms) Archaea and Bacteria

Members of Domains (Kingdoms) Archaea and Bacteria usually can be readily identified under a microscope because they are single-celled and *prokaryotic.* Prokaryotic cells have no nuclei or other organelles bounded by membranes. Bacteria occur in three basic forms: *cocci,* which are more or less spherical; *bacilli,* which tend to be rod-shaped; and *spirilli,* whose cells are twisted like corkscrews. The kingdom previously known as Monera is divided into two kingdoms, based on some fundamental differences in the chemistry, metabolism, and RNA molecules of the cells. Domain (Kingdom) Archaea includes anaerobic *methane bacteria, salt bacteria* that carry on a simple form of photosynthesis and live in water saturated with salt; and *sulfolobus bacteria,* which live exclusively in hot springs. Domain (Kingdom) Bacteria includes the majority of the better-known bacteria, most of which are *saprobes* that depend on nonliving organic matter for their energy sources; *parasites* that use other living organisms as their energy sources; and *autotrophic bacteria* that synthesize organic compounds from inorganic compounds by photosynthesis or other means.

Nearly all bacterial cells are considerably smaller than those of complex plants and animals and are best examined with the highest power of a compound microscope. We will examine representative bacteria that have no pigments within their cells and cyanobacteria that have chlorophyll and other pigments in membranes within the cells.

Kingdom Protista

Members of Kingdom Protista all have *eukaryotic* cells with nuclei and various organelles discussed in earlier exercises. With the exception of *protozoans, sponges, water molds,* and *slime molds* (slime molds are the only group of these organisms that will be discussed here), virtually all members of Kingdom Protista possess chlorophyll and other pigments confined to chloroplasts.

The active state of *slime molds* is called a *plasmodium*. Unlike a true mold, which consists of delicate threads that in most fungi are compartmentalized into individual cells, a plasmodium consists of a multinucleate mass of cytoplasm without cell walls. Plasmodia move over dead leaves and debris in a "crawling-flowing" motion. As the plasmodia move, they engulf bacteria and other food materials. Slime molds, like fungi, have glycogen as a primary food reserve and other fungus-like features such as stationary reproductive bodies (*sporangia*). They differ sharply from true fungi, however, in their flagellated reproductive cells— a feature that suggests they originated from other members of Kingdom Protista.

In this exercise, you will also be introduced to a few representatives of the thousands of species of pigmented *algae*. Algae vary in size from minute single-celled organisms to giant kelps that may attain lengths of 45 meters (nearly 150 feet). External water is essential to algae completing their life cycles, and the great majority of them are aquatic. All possess chlorophyll *a*, but each phylum exhibits unique combinations of pigments, different food reserves, and distinctive reproductive cells. They occur in nature as *single cells, colonies, filaments,* or *thalli* (flattened bodies), or in mutually beneficial associations with fungi.

A. Non-Photosynthetic Bacteria

Examine the plates of bacterial colonies growing on *agar,* a gelatinlike substance obtained from several red and a few brown seaweeds. Note the colors and textures of the colonies, which consist of many thousands of bacteria. Mount a *small* amount of bacteria in a drop of water on a slide by touching the tip of a probe to a colony and then vigorously rotating it in the drop of water. Cover with a coverslip, and, after locating cells under low power, switch to high power. Note the shapes and sizes of the bacteria.

In the 19th century, Christian Gram discovered that some bacteria retain a stain he devised and others do not retain it. His stain became known as the *Gram stain;* those bacteria retaining the stain were called *gram-positive,* and those not retaining the stain were called *gram-negative.* Variations of the Gram stain are now routinely used as a first step in identifying bacteria. If your instructor decides to have you check bacteria provided for their response to a Gram stain, he/she will show you how to make bacterial *smears.* You should then be ready to proceed as follows.

On *clean* microscope slides, make smears of bacteria in the plates marked **A** and **B,** one smear to a slide. Dry the slides by passing them *rapidly* through a Bunsen burner flame four or five times. After the slides are dry, add a drop of gentian (crystal) violet dye to each slide. Tilt the slides so that any excess dye drains away from the bacteria, and add a drop of Gram's iodine reagent. Allow the iodine to stand for 1 to 2 minutes, and then add, one at a time, drops of 95% ethyl alcohol until the violet color is no longer apparent to the naked eye. Now add one drop of safranin O dye and wait for 30 seconds. Then wash *gently* with water, add a coverslip, and examine with the microscope. Note that the bacteria on one slide are stained purple and those on the other are not. The *gram-positive* bacteria are stained purple; the *gram-negative* are not stained.

B. Cyanobacteria

Mount a *small* amount of the cyanobacteria *Nostoc* or *Anabaena* on a slide, or, if fresh material is not available, examine the prepared slides of these cyanobacteria. Note that the pigments are diffused throughout the cells and not located in plastids. Can you see any nuclei? Are there any colorless *heterocysts* (cells that appear to have a slightly thicker wall) scattered throughout the filaments? Heterocysts are nitrogen-fixing cells at which cyanobacterial filaments may fragment (break). Are there any dense-looking cells that are somewhat oblong in outline (*akinetes*) present at the ends or within the filaments? Akinetes are resistant to freezing and desiccation and are a means of ensuring the survival of the organisms over winter or when water is lacking.

C. Pond Water Organisms

Agitate the pond water and place a single drop on a *clean* slide; add a coverslip. Examine with the compound microscope. Make drawings or diagrams of at least three different organisms. To help you identify the organisms, there are pictures of some of the more common ones on the pages at the end of this exercise. Your instructor may help you identify other organisms not illustrated. Do you see any *motile* forms (i.e., forms that are moving)? Movement may be by means of whiplike tails called *flagella* or by means of numerous short, moving hairs called *cilia. Diatoms* have a rigid, glasslike cell wall composed primarily of silica; their movement may be brought about by means of cytoplasm extending through pores and functioning somewhat like a Caterpillar tractor track. Both flagella and cilia are minute in diameter and may be difficult to see without special equipment or techniques. Notice the wide variety of chloroplast types and the small, round, colorless *pyrenoids* (starch accumulation centers found in green algae) on some of the larger chloroplasts (they may or may not be present on your particular slide).

D. *Spirogyra*

Mount a small amount of *Spirogyra* in a drop of water on a slide. Locate a pair of *conjugating filaments.* If your fresh material is not conjugating, examine a prepared slide showing this. Observe the *papillae* that unite, forming *conjugation tubes* (short cylindrical tubes between adjacent cells). Are any *gametes* migrating through the conjugation tubes to adjacent cells? If the cells of one filament are empty, note the relatively thick *zygotes* in the cells opposite those of the empty filament. No special asexual reproductive cells are produced by *Spirogyra.* Instead, new cells are added by mitosis after filaments break.

E. *Ulothrix* and *Oedogonium*

Examine the cultures of *Ulothrix* and *Oedogonium* available, and study the prepared slides of these two green algae. How do these two algae differ from one another and from *Spirogyra* with respect to *chloroplasts* and *reproduction?*

F. Diatomaceous Earth

In a drop of water on a clean slide, mount a *small* amount of *diatomaceous earth*, which consists of the "shells" of millions of marine diatoms. What are the most common shapes of marine diatoms? Is there variation in the patterns of pores present? Note that diatoms may appear to have one shape in *valve* (top or bottom) view and another shape in *girdle* (side) view.

G. Dinoflagellates

In a drop of water on a clean slide, mount a *small* amount of the dinoflagellate material provided, or, alternatively, examine the demonstration that has been set up. Note the "armor" plates. How are the grooves in which the flagella are located arranged?

H. Seaweeds

Examine the herbarium specimens of marine algae (seaweeds) on display. Although some of the larger seaweeds have forms of food- or water-conducting tissues, none have true xylem or phloem, root, or leaves. Do any of them have *holdfasts* (rootlike structures that anchor them to rocks), *bladders* (bulblike swellings that enable seaweeds to float), *stipes* (stalks), or *blades* (flattened, leaflike bodies)?

I. Slime Molds

With the aid of your dissecting microscope, examine the petri dishes with living *plasmodia* of *slime molds*. Focus on the leading edge of a plasmodium, and note the rapid flowing of the protoplasm. Does the direction of flow ever change? Can you see individual cells?

Examine the specimens of slime mold reproductive bodies available. If expendable materials are provided, mount a *sporangium* in a drop of water on a slide. Observe the numerous spherical *spores* and the *capillitial threads*

interspersed among the spores. A few species of slime molds lack capillitial threads, which are unknown in true fungi.

Drawings to Be Submitted

1. Draw a group of bacterial cells from those you mounted in water. Also draw one or two filaments of cyanobacteria such as *Nostoc* or *Anabaena*. Be sure to indicate the magnifications of your drawings.
2. Draw at least three different algae occurring in your pond water. If they are not illustrated on the pages at the end of this exercise, ask your instructor to identify them for you.
3. Label the drawings of *Spirogyra* and *Ulothrix* provided. Labels for *Spirogyra* should include VEGETATIVE FILAMENT, PYRENOID, CHLOROPLAST, CONJUGATING FILAMENTS, PAPILLAE, GAMETE, ZYGOTE, and GERMINATING ZYGOSPORE. Labels for *Ulothrix* should include HOLDFAST, ZOOSPORES, NEW FILAMENT, MATURE FILAMENT, GAMETES, FERTILIZATION, ZYGOTE, and MEIOSIS. Show where MEIOSIS takes place in both organisms.
4. Draw a filament of *Oedogonium* from a prepared slide. Label VEGETATIVE CELL, OOGONIUM, EGG, ANTHERIDIUM, and SPERM.
5. Draw a diatom, showing its markings.
6. Draw a dinoflagellate showing its grooves and "armor" plates.
7. Draw a marine alga (seaweed). Identify any HOLDFASTS, BLADDERS, STIPES, or BLADES present.
8. With the aid of your dissecting microscope, draw several slime mold sporangia. Label SPORANGIUM and CAPILLITIAL THREADS.

When you have completed your assignments for this exercise, break off a portion of bread small enough to fit within the petri dish provided. Then add **no more than *one* drop** of water to the bread (if you add more than one drop, yeasts are very likely to multiply and interfere with the growth of other fungi you will be trying to cultivate). Next, sprinkle dust from the corners of the floor in the laboratory or elsewhere, or comb your hair over the bread. Close the dish and print your name and section number on the outside, then set it aside until the next laboratory session. At that time, we will examine any fungi that develop on the bread.

CYANOBACTERIA

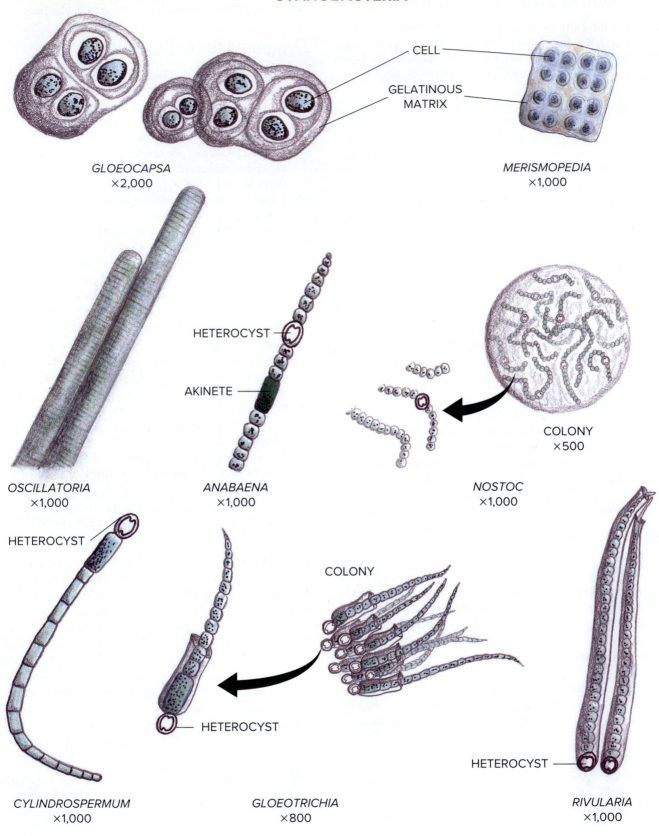

GLOEOCAPSA
×2,000

CELL

GELATINOUS
MATRIX

MERISMOPEDIA
×1,000

OSCILLATORIA
×1,000

HETEROCYST

AKINETE

ANABAENA
×1,000

COLONY
×500

NOSTOC
×1,000

HETEROCYST

CYLINDROSPERMUM
×1,000

COLONY

HETEROCYST

GLOEOTRICHIA
×800

HETEROCYST

RIVULARIA
×1,000

GREEN ALGAE AND EUGLENA

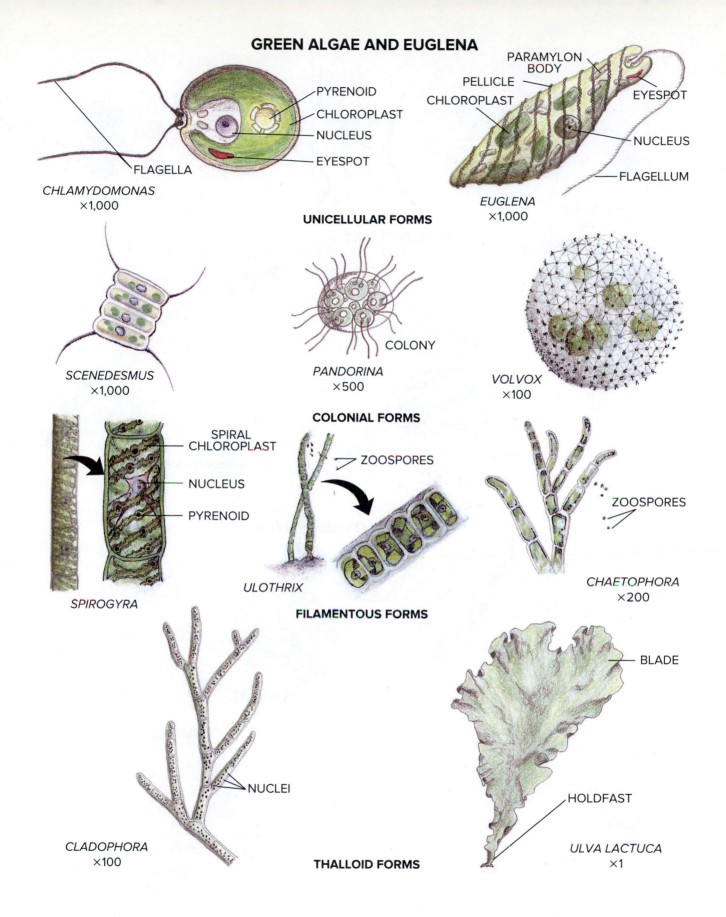

CHLAMYDOMONAS
×1,000

PYRENOID
CHLOROPLAST
NUCLEUS
EYESPOT
FLAGELLA

PARAMYLON BODY
PELLICLE
CHLOROPLAST
EYESPOT
NUCLEUS
FLAGELLUM

EUGLENA
×1,000

UNICELLULAR FORMS

SCENEDESMUS
×1,000

PANDORINA
×500
COLONY

VOLVOX
×100

COLONIAL FORMS

SPIRAL CHLOROPLAST
NUCLEUS
PYRENOID

SPIROGYRA

ZOOSPORES

ULOTHRIX

ZOOSPORES

CHAETOPHORA
×200

FILAMENTOUS FORMS

NUCLEI

CLADOPHORA
×100

BLADE

HOLDFAST

ULVA LACTUCA
×1

THALLOID FORMS

OTHER GREEN ALGAE

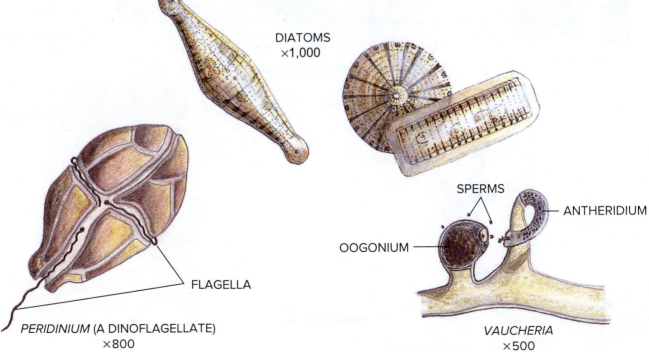

TRIBONEMA
×500

ANKISTRODESMUS
×500

PROTOCOCCUS
×500

ZYGNEMA
×500

MICRASTERIAS
×200

NUCLEUS

VACUOLE

CLOSTERIUM (A DESMID)
×200

ALGAE OF OTHER PHYLA

DIATOMS
×1,000

SPERMS

ANTHERIDIUM

OOGONIUM

FLAGELLA

PERIDINIUM (A DINOFLAGELLATE)
×800

VAUCHERIA
×500

Exercise 14

NAME _____

LAB SECTION NO. _____

DATE _____

GREEN ALGAE

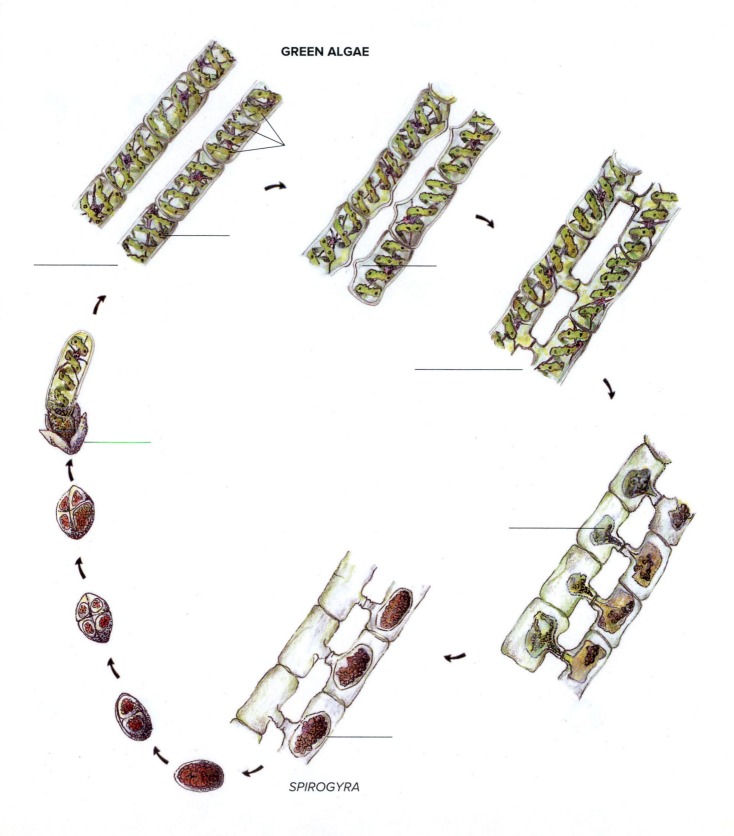

SPIROGYRA

GREEN ALGAE

ASEXUAL

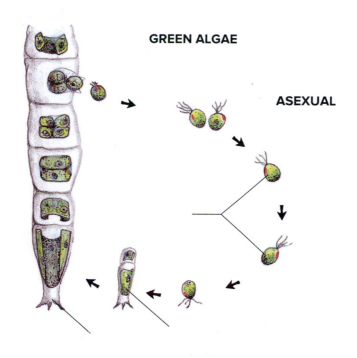

SEXUAL

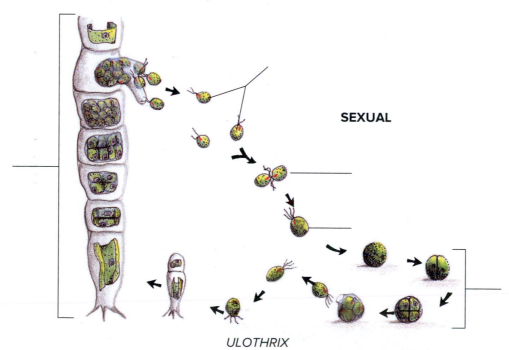

ULOTHRIX

Review Questions 14

NAME _____

LAB SECTION NO. _____

DATE _____

1. What is the difference between *prokaryotic* and *eukaryotic* cells? _____

    _____

2. If a bacterium is *gram-positive,* how does it respond to a Gram stain? _____

3. Which organisms are most likely to contain *heterocysts?* _____

    _____

4. Of what advantage to a cyanobacterium is an *akinete?* _____

    _____

5. What is the function of *conjugation tubes?* _____

    _____

6. Specifically where in a cell are *pyrenoids* located? _____

7. What substance gives rigidity to *diatom* cell walls? _____

8. How do diatoms move? _____

    _____

    How do dinoflagellates move? _____

9. How does the *plasmodium* of a slime mold differ from the *mycelium* of a true fungus? _____

    _____

10. What is the equivalent of a *root* in a seaweed? _____

Laboratory Preparation Quiz 14

NAME _____

LAB SECTION NO. _____

DATE _____

Domains (Kingdoms) Archaea and Bacteria; Kingdom Protista

1. What is a *pyrenoid?* _____

2. In which alga would you expect to find *conjugation tubes?* _____

3. If you find a pair of algal filaments conjugated but the cells of one of them are empty, what are the dark objects in the

   cells of the other filament? _____

4. What are the colorless, slightly thicker-walled cells present in some cyanobacteria? _____

5. What is an *akinete?* _____

6. In addition to differences in the way they reproduce, what should help you distinguish *Spirogyra* from *Oedogonium?*

   _____

7. What is a *motile* alga? _____

8. What type of organism mentioned in this exercise may have *bladders?* _____

   _____

9. Where are the pigments of cyanobacteria located? _____

10. Which algae have rigid, glasslike walls? _____

Kingdom Fungi (Mycota)

Materials

1. Preserved, dried, or living specimens of puffballs, stinkhorns, earth stars, rusts, smuts, bracket fungi, bird's-nest fungi, and any other available fungi
2. Prepared slides of *Rhizopus, Penicillium, Peziza, Coprinus,* and *Physcia*
3. Examples of crustose, foliose, and fruticose lichens

Some Suggested Learning Goals

1. Know the sexual life cycle of the black bread mold *Rhizopus* and how it reproduces asexually.
2. Learn the various parts of the reproductive structures of a *sac fungus.*
3. Know the life cycle of a common *mushroom,* and be able to explain what occurs in the various phases.
4. Understand how the production of asexual *spores* in *Penicillium* differs from the production of asexual spores in *Rhizopus.*
5. Understand the nature and structure of a *lichen,* the three basic forms of lichens, and how lichens reproduce.

Introduction

Kingdom Fungi, also called Kingdom Mycota, includes thousands of organisms such as molds, chytrids, mildews, mushrooms, puffballs, smuts, rusts, shelf or bracket fungi, jelly fungi, bird's-nest fungi, and stinkhorns. Unlike many members of Kingdom Protista, no fungus has chlorophyll, and they are dependent on other organisms for their food. Some are *saprobes* (organisms that live on dead organic matter); others are *parasites* (organisms that live at the expense of living host organisms); while others may be mutualistic *symbionts* (both species benefit from each other). *Lichens* are examples of organisms thought to involve symbiotic relationships, where one of the components is a fungus and the other an alga.

A. True Fungi

1. Note the living and preserved fungi on display. With the aid of your textbook, assign them to their proper *phyla.*
2. Examine the dishes containing the black bread mold *Rhizopus.* If you focus carefully with your dissecting microscope, you will see upright hyphae (*sporangiophores*) with somewhat pebbly, spherical

sporangia at the tips. While they are developing, the *spores* in the sporangia are white, but as they mature, they turn black.

Turn now to the prepared slide of *Rhizopus.* This slide shows both asexual and sexual reproductive stages. Find the larger black *zygospores* that developed after *gamete nuclei* united. Note the *suspensors,* one on either side of a zygospore. One suspensor is often larger than the other. Also find a sporangium. Note the spores and the dome-shaped *columella* within each sporangium. Notice that the hyphae are *coenocytic* (i.e., there are no crosswalls within the tubelike hyphal threads). You may be able to see rootlike hyphae, called *rhizoids,* anchoring the fungus.

3. Examine the prepared slide of *Penicillium* with the high power of your compound microscope. If you focus carefully around the edge of the material, you should be able to detect more or less parallel rows of *spores* being pinched off from *conidia,* the name given to *hyphae* that give rise to asexual *spores.*

4. Examine the prepared slide of a *Peziza* cup (*ascoma*) with the *low power* of your compound microscope. Notice that the cup has been formed from densely interwoven hyphae. Many of the hyphae forming the outer part of the ascoma have cells with a single nucleus (such hyphae are said to be *monokaryotic*). Other hyphae with cells having two nuclei (*dikaryotic* hyphae) are interspersed with the monokaryotic hyphae, especially in the central part of the ascoma. The two nuclei in the cells at the tips of the dikaryotic hyphae unite and undergo meiosis, producing a neat layer of upright, tubular *asci* (singular: *ascus*) at the open end of the cup. The layer of asci is called a *hymenium.* Turn now to high power, and count the *ascospores* in each ascus. Eight ascospores are normally produced in each ascus.

5. Examine the prepared slide of a cross section through the *cap* of a common mushroom (*Coprinus*) with the low power of your compound microscope. Note the *gills* radiating out like the spokes of a wheel from the *stipe* (stalk). Turn now to high power, and examine part of a gill. Note the small, club-shaped *basidia* lining either side of the gill. Can you detect any tiny, peglike *sterigmata* at the tips of the basidia? The darker *basidiospores* are formed after a *zygote nucleus* has undergone *meiosis.* Zygote nuclei are formed by the union of the two nuclei in each cell of a *dikaryotic*

mycelium. A dikaryotic mycelium is initiated when two *monokaryotic mycelia* come in contact and a cell of one mycelium unites with a cell of the other. Each monokaryotic cell contains one nucleus that does not unite with the other nucleus in the new cell. The new cell and its nuclei then divide in a special way so that each daughter cell has one nucleus from each of the "parent" mycelia. The dikaryotic mycelia that develop eventually may form *buttons* that become *mushrooms* (*basidiomata*). You will not be able to discern zygote nuclei or the two kinds of mycelia on your slide.

B. Lichens

A lichen is a mutually beneficial combination of an alga and a fungus in an intimate association. The vast majority of lichens consist of a single species of *ascomycete* (sac fungus) and an alga—usually a green alga. Each species of lichen has a fungus unique to that species, but a species of alga may occur in several different lichens. The alga, through photosynthesis, furnishes the food, while the fungus apparently supplies chemical substances that promote the alga's growth and also helps to retain moisture, at least during part of the year, for the alga's needs. On the basis of growth habit, lichens are grouped into *crustose* (forming a fine crusty growth), *foliose* (forming a somewhat leaf-like mat), and *fruticose* (having erect or pendent growth) species (Fig. 15.1). When Latin names are assigned to them, they are classified on the basis of the fungus present. Examine the slide of *Physcia* or other lichen *thallus* (flattened body) available. Distinguish between the algal cells and the fungal hyphae.

Drawings to Be Submitted

1. Label the accompanying drawings of *Rhizopus.* Labels should include GAMETES, SUSPENSOR(S), ZYGOSPORE, GERMINATING ZYGOSPORE, SPORANGIUM, COLUMELLA, SPORE(S), SPORANGIOPHORE, and RHIZOID(S).
2. Draw a portion of the cup of the sac fungus, *Peziza.* Label ASCOMA, ASCUS, ASCOSPORE(S), and HYPHAE.
3. Draw part of a *Penicillium* mycelium. Label CONIDIUM, SPORES, and HYPHAE. (You will need to turn to high power for this drawing.)
4. Label fully the accompanying drawings of a mushroom in the various phases of its life cycle. Include STIPE (stalk), GILL, CAP, BASIDIOSPORE, STERIGMATA, BASIDIUM, ZYGOTE NUCLEUS, MONOKARYOTIC MYCELIUM, and DIKARYOTIC MYCELIUM. Also, indicate where MEIOSIS occurs.

(a)

(b)

(c)

FIGURE 15.1   THREE TYPES OF LICHENS: CRUSTOSE (TOP), FOLIOSE (MIDDLE), AND FRUTICOSE (BOTTOM).
© Kingsley Stern

5. Draw any one or two of the larger fungi on display. If present, label CAP, STIPE (stalk), and GILL.
6. Draw a portion of a lichen thallus as seen with the aid of the low power of your compound microscope. Label ASCOMA, ASCI, HYPHAE, and ALGAE.

NAME _____

LAB SECTION NO. _____

DATE _____

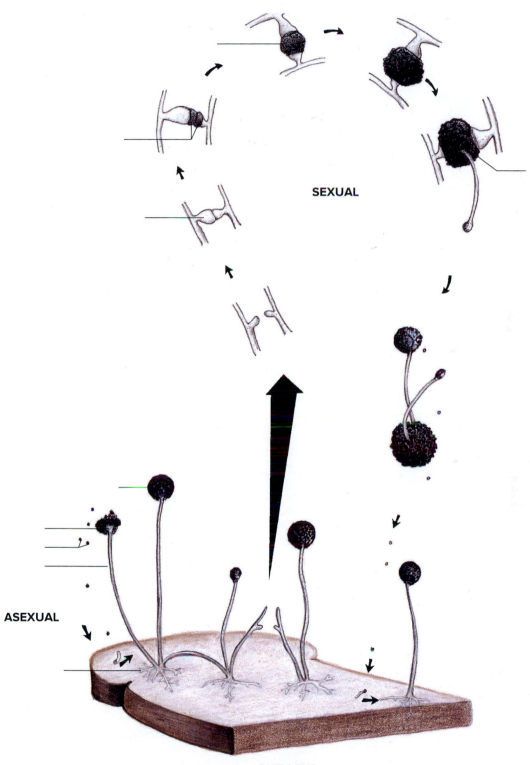

SEXUAL

ASEXUAL

RHIZOPUS

Exercise 15

NAME _____

LAB SECTION NO. _____

DATE _____

KINGDOM FUNGI AND LICHENS

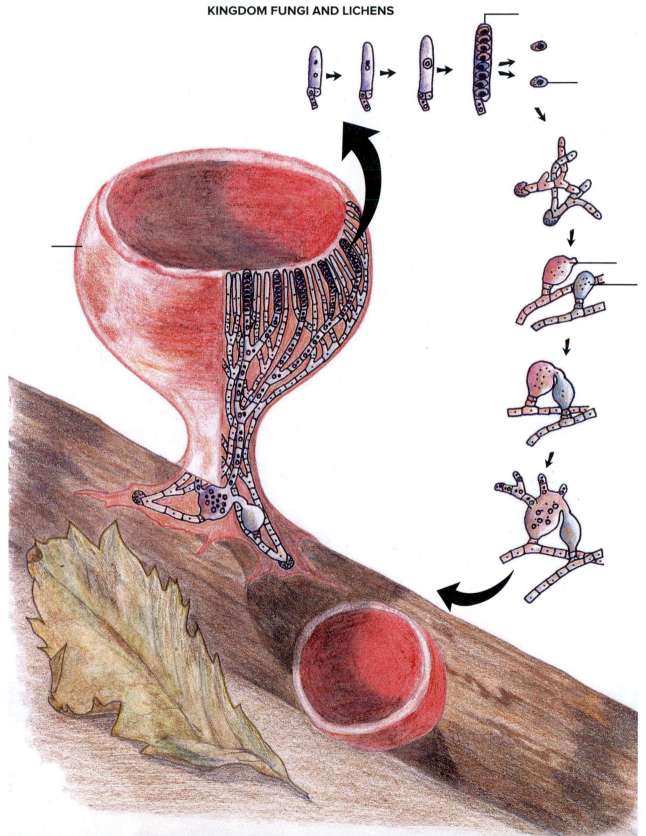

SAC FUNGUS (*ASCOMYCETE*)

Exercise 15

SMUT

PUFFBALL

JELLY FUNGUS

STINKHORN

BRACKET FUNGUS

EARTH STAR

RUST

BIRD'S-NEST FUNGUS

Exercise 15

NAME _____

LAB SECTION NO. _____

DATE _____

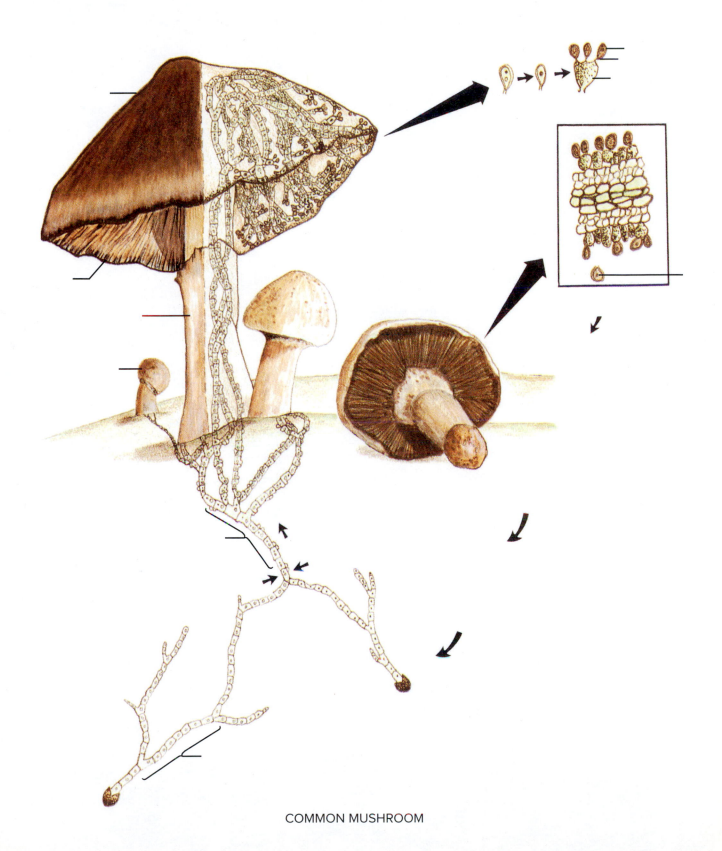

COMMON MUSHROOM

Review Questions 15

NAME _____

LAB SECTION NO. _____

DATE _____

1. How does the *plasmodium* of a slime mold differ from the *mycelium* of a true fungus? _____

   _____

2. What is the difference between *capillitial threads* and a *columella?* _____

   _____

3. If you had a powerful microscope and *hyphae* of a bread mold (e.g., *Rhizopus*) and *hyphae* of a common mushroom, both stained so that you could see the parts clearly, how could you tell them apart? _____

   _____

4. How are *spores* produced in *Rhizopus?* _____

   In *Penicillium?* _____

5. What is a *hymenium?* _____

6. How many *ascospores* are usually produced in each *ascus?* _____

7. What club-shaped reproductive structures are produced on either side of a mushroom *gill?* _____

   _____

8. How is a *dikaryotic* mycelium initiated? _____

9. What is the function of an alga in a *lichen?* _____

10. Which category of lichen *thallus* either stands up or hangs down? _____

   _____

Laboratory Preparation Quiz 15

NAME _____

LAB SECTION NO. _____

DATE _____

Kingdom Fungi (Mycota)

1. What is the active phase of a *slime mold* called? _____

2. In addition to *spores,* what would you expect to see inside a slime mold *sporangium?* _____

   _____

3. What is a *saprobe?* _____

4. What structures are found on either side of the *zygote* in *Rhizopus?* _____

   _____

5. Would you expect to find *crosswalls* in the hyphae of *Rhizopus* (black bread mold)? _____

   _____

6. What do we call the tubes with single files of spores that are found in a row in the cup (*ascoma*) of *Peziza?* _____

   _____

7. In a *mushroom,* what are the club-shaped structures that bear spores? _____

   _____

8. To which *phylum* do the algae most commonly found in lichens belong? _____

   _____

9. Which type of lichen forms a somewhat leaflike mat? _____

10. What is the body of a lichen called? _____

Kingdom Plantae: Bryophytes and Ferns

16

Materials

1. Live mosses with sporophytes attached
2. Prepared slide of moss protonema
3. Live *Marchantia* with archegoniophores and antheridiophores
4. Petri dish with live protonemata (demonstration)
5. Prepared slides of longitudinal sections of archegonial heads of *Mnium* (or similar moss)
6. Live hornworts (demonstration)
7. Variety of live fern plants, one with expendable fronds that have mature sori
8. Live prothalli (demonstration)
9. Prepared slides (whole mounts) of bisexual prothalli

Some Suggested Learning Goals

1. Understand how the development of *gametangia* (structures in which sex cells are produced) and *zygotes* of members of the Plant Kingdom differs from the development of gametangia and zygotes in members of other kingdoms.
2. Understand how the form and structure of *bryophytes* differs from that of more complex plants.
3. Know what develops or takes place in each phase of the life cycle of a *moss*.
4. Know what develops or takes place in each phase of the life cycle of a *liverwort*.
5. Learn how asexual and sexual reproduction of both *thalloid* and *"leafy"* liverworts differs from that of mosses.
6. Be able to explain basic differences between the sporophytes of *ferns* and *mosses*.
7. Know the life cycle of a typical *fern*.
8. Understand the nature of a *prothallus* and a *sorus* and the roles they play in a fern life cycle.

Introduction

The Plant Kingdom, as it is understood today, includes bryophytes and lower vascular plants such as ferns, as well as gymnosperms and angiosperms.

In algae, fungi, and other relatively primitive organisms that in the past were regarded as plants, the *gametes* (sex cells) are produced in single-celled *gametangia,* and the *zygote* often undergoes meiosis directly. Beginning with the *bryophytes* (e.g., mosses), the gametes are produced in gametangia that are composed of many cells; the zygote,

through mitosis, develops into an *embryo* that, in turn, develops into a diploid *sporophyte. Spores* are produced by *meiosis* within a specialized part of the *sporophyte.*

Alternation of Generations in bryophytes and ferns is marked by the development of distinct, separate gametophyte and sporophyte bodies. In bryophytes, the sporophyte, while a distinct body in itself, is dependent on the gametophyte for most of its nutrition. In ferns, however, both the gametophyte and the sporophyte are photosynthetic and they are independent of each other.

A. Bryophytes (Phyla Bryophyta, Hepaticophyta, and Anthocerotophyta)

Mosses, liverworts, and *hornworts* are included in these three phyla. Bryophytes differ from higher plants in lacking xylem and phloem, although some do have specialized cells that can conduct a little water and food in solution. Some species may form extensive low mats consisting of dozens or even hundreds of plants. Because true xylem and phloem are lacking, however, none of the individual plants become very large. They cannot grow or function very long without external moisture, hence their usual association with damp habitats.

Examine the clump of moss provided. The clump consists of small green "leafy" *gametophyte* plants. ("Leafy" is in quotation marks because unlike true leaves, which are diploid, those of mosses consist of a single layer of haploid cells; moss and liverwort "leaves" also have no internal structure or stomata.) The "leaves" do, however, carry on photosynthesis like the true leaves of more complex plants.

Some of the moss plants may have a thin stalk, or *seta,* emerging from the tip. A *capsule (sporangium)* develops at the free end of each seta. The seta and capsule are diploid ($2n$) and constitute the *sporophyte. Sporocytes* (not visible here) within each sporangium undergo *meiosis,* producing *spores.* The sporangium is usually partially to completely covered with a "pixie cap" called a *calyptra.* The calyptra develops from archegonial tissues and is therefore n (*haploid*). When the calyptra is removed, a second, smaller ($2n$) cap may be seen covering the free end of the capsule. This smaller cap, the *operculum,* develops with the sporophyte and eventually pops off, allowing the spores to disperse. Release of the spores is partially controlled by tiny *peristome teeth* at the rim of the capsule; the peristome teeth, which resemble tiny, cross-ribbed shark's teeth, move in response to changes in humidity.

Turn now to a prepared slide labeled "Moss protonema." A *protonema* is an algalike body that develops when a moss spore germinates. Notice the chloroplasts present in each cell and that the transverse walls of the cells usually are not strictly at right angles to the other walls. Note, also, the "*buds*" that are developing along some of the threads. These buds become new "leafy" gametophyte plants. Some may already have rootlike *rhizoids* at their bases. Rhizoids are only one cell thick; they may anchor bryophyte plants in the same way true roots do, but like the remainder of a moss gametophyte, they have no xylem or phloem and can absorb water slowly only in very limited amounts. The word *rhizoid* should not be confused with *rhizome,* which is a term applied to the horizontal stems of ferns and higher plants.

Next turn to a slide of moss *archegonia*. Archegonia are female reproductive structures of mosses produced at the tips of female gametophyte plants. Occasionally both male and female reproductive structures are produced on the same plant. Each archegonium loosely resembles a tiny vase with a narrow neck, the enlarged base itself being elevated on a short, relatively wide stalk. Although there are usually several to many archegonia produced at the tip of each plant, the archegonia are not always strictly upright, and they are interspersed among sterile, multicellular hairs, called *paraphyses*. When microscope slides of moss archegonia are made, very thin longitudinal sections are cut and stained. Parts of the archegonia and paraphyses are often sliced off; because of this you may not have a complete archegonium on your slide. You should, however, be able to see at least one archegonium with its base intact. The cavity within the archegonium base contains an *egg*.

Turn now to a slide of moss *antheridia*. These structures, before they are sliced, are shaped like miniature clubs; they contain numerous *sperms*. Paraphyses usually are present among the antheridia. In nature, a sperm swims down the neck of an archegonium and unites with the egg, forming a *zygote*. As the zygote divides, it forms an *embryo,* which is dependent on the gametophyte for its nutrition. The embryo then develops into a *sporophyte,* consisting of a seta and capsule. Even the mature sporophyte is still largely dependent on the gametophyte for its energy.

Liverworts have nearly all of the reproductive structures found in mosses. Although there are many species of "leafy" liverworts, some of the most common and best-known forms are *thalloid*. Thalloid liverworts have flattened bodies that look a little like bright-green foliose lichens. Examine the thalloid liverworts provided. Some, such as *Marchantia,* have their archegonia and antheridia elevated above the *thallus* on umbrella-like *archegoniophores* and disc-shaped *antheridiophores*. Many thalloid liverworts also reproduce asexually by means of *gemmae,* which are tiny, lens-shaped pieces of vegetation produced within *gemmae cups*. The gemmae cups are scattered over the surface of the thallus. Each gemma is potentially capable of developing into a new thallus.

Examine the demonstration of *hornworts* provided. How do the sporophytes of hornworts differ in appearance from those of liverworts and mosses? Are there any other apparent features that distinguish hornworts from other bryophytes?

B. Ferns (Phylum Polypodiophyta)

Unlike the bryophytes, the ferns do possess true conducting tissues (xylem and phloem), and the sporophyte is the more conspicuous phase of the life cycle.

Examine the fern plants on display. The leaves, or *fronds,* arise from a horizontal stem (*rhizome*). Notice the small brownish patches on the backs of mature fronds. Remove a small part of a frond that has these patches and examine them with the aid of your dissecting microscope. Each discrete patch is called a *sorus* and consists of a cluster of *sporangia*. The sporangia are often partially or wholly covered by a transparent, umbrella-like *indusium*. *Sporocytes* within the sporangia undergo meiosis, producing *spores*. The spores are released through the springlike action of the *annulus,* which is composed of heavy-walled cells around most of the edge of the sporangium.

Now examine the green, heart-shaped *prothalli* that constitute the *gametophytes* of ferns. Both living and preserved prothalli may be provided. Some prothalli produce only *archegonia,* others only *antheridia*. The prothallus on the microscope slide produces both. Find an antheridium, often located among the rootlike *rhizoids*. It is circular in outline and contains *sperms*. Then find an archegonium, which is roughly the same size as an antheridium but has a short neck. Archegonia each contain a single *egg;* they often tend to be close to the notch of the prothallus at the top. In nature, a sperm unites with the egg in an archegonium, and the *zygote* develops into a new *sporophyte,* with which our study of ferns began.

Drawings to Be Submitted

1. Label the following on the drawings of the moss life cycle provided: MALE GAMETOPHYTE, FEMALE GAMETOPHYTE, EGG, ANTHERIDIUM, SPERM, ARCHEGONIUM, ZYGOTE, DEVELOPING SPOROPHYTE, EMBRYO, MATURE SPOROPHYTE, SPOROCYTE, CAPSULE, PERISTOME, OPERCULUM, CALYPTRA, SPORES, PROTONEMA, BUD, and RHIZOIDS. Also indicate where MEIOSIS occurs.

2. Draw portions of sections through the tips of moss gametophytes from prepared microscope slides, using the low power of your compound microscope. Show at least one good ARCHEGONIUM and several PARAPHYSES in the FEMALE GAMETOPHYTE and at least one or two ANTHERIDIA and several PARAPHYSES in the MALE GAMETOPHYTE. Also label EGG and SPERM(S). Be sure to reread the

comments in Part A about how parts of structures may be cut off during the manufacture of the slides.

3. Draw HABIT SKETCHES (i.e., how the organisms appear in nature) of a thalloid liverwort and a hornwort.

4. Draw a fern SORUS, with the aid of the highest power of your dissecting microscope. Label SPORANGIA (and INDUSIUM, if present).

5. Label the following on the drawings of the fern life cycle provided: DEVELOPING GAMETOPHYTE (IMMATURE PROTHALLUS), PROTHALLUS, RHIZOIDS, ARCHEGONIUM, EGG, ANTHERIDIUM, SPERM, ZYGOTE, EMBRYO, YOUNG SPOROPHYTE, SPOROPHYTE, FROND, ROOTS, RHIZOME, SORUS, SPORANGIUM, SPOROCYTES, and SPORES. Also indicate where MEIOSIS occurs.

6. Draw a fern prothallus from the prepared slide provided. Label ARCHEGONIUM, ANTHERIDIUM, and RHIZOIDS. (Use the lowest power of your compound microscope.)

7. Draw a fern frond, showing the position of the SORI.

NAME _____

LAB SECTION NO. _____

DATE _____

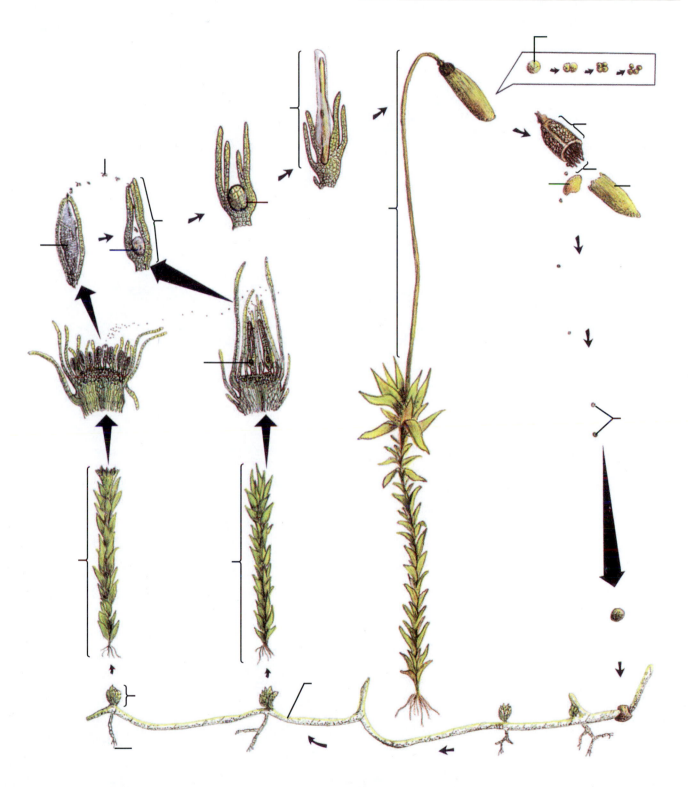

MOSS LIFE CYCLE

NAME

LAB SECTION NO.

DATE

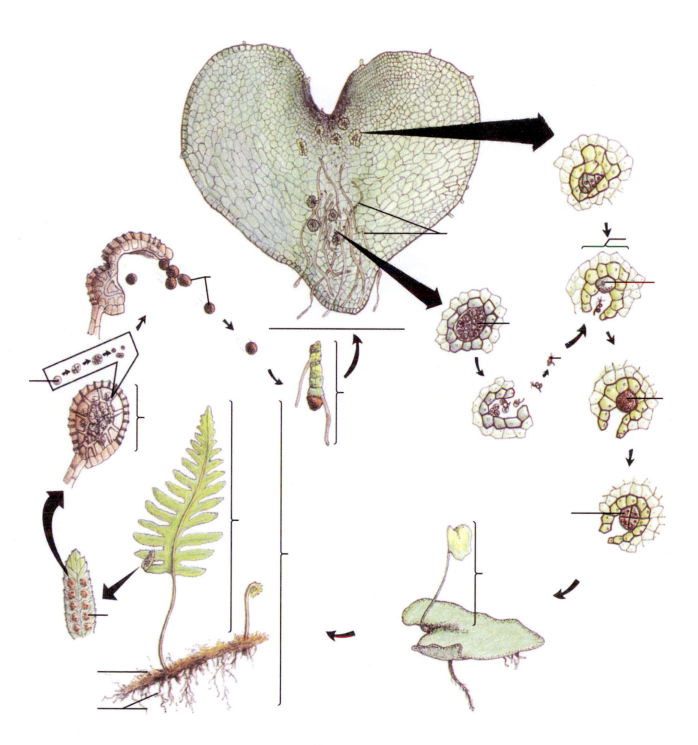

FERN LIFE CYCLE

Review Questions 16

NAME _____

LAB SECTION NO. _____

DATE _____

1. How do moss "leaves" differ from the leaves of more complex plants? _____

_____

2. What is the difference between a *calyptra* and an *operculum?* _____

_____

3. How is the release of spores controlled in mosses? _____

_____

4. Where does *meiosis* take place in mosses? _____

5. Where, in mosses, are *zygotes* and *embryos* formed? _____

6. In *Marchantia,* what is the function of *archegoniophores* and *antheridiophores?* _____

_____

7. What are all the parts of a complete *sorus?* _____

8. Where, specifically, are fern *antheridia* located? _____

9. What parts of a fern are diploid? _____

Where, in a fern, does the switch from diploid to haploid take place? _____

Where, in a fern, does the switch from haploid to diploid take place? _____

Laboratory Preparation Quiz 16

NAME _____

LAB SECTION NO. _____

DATE _____

Kingdom Plantae: Bryophytes and Ferns

1. In addition to seeds, what do higher plants have that *bryophytes* lack? _____
   _____

2. Which phase in the life cycle of a moss consists of a "leafy" plant? _____
   _____

3. In which specific structure of a moss are *sperms* produced? _____

4. What is the toothed structure in a moss *sporophyte* that controls the release of spores from a *sporangium?* _____
   _____

5. How does a *thalloid liverwort* differ in appearance from a *moss?* _____
   _____

6. What is a cluster of fern sporangia called? _____

7. Where does *meiosis* take place in ferns? _____

8. What name is applied to the *gametophyte* of ferns? _____

9. Where are fern *antheridia* produced (i.e., among what structures on the gametophyte)? _____
   _____

10. What are the differences among *rhizoids, roots,* and *rhizomes?* _____
    _____
    _____

Kingdom Plantae: Gymnosperms

© Kingsley Stern

Materials

1. Fresh pine branch with cluster of pollen cones (demonstration)
2. Fresh pine, fir, and other conifer branches for examination of leaves
3. Pine seed cones with seeds on the scales
4. Conifer pollen
5. Prepared slides of longitudinal sections through a pine ovule
6. Demonstrations of specimens of cycads, *Gnetum*, *Ginkgo*, *Ephedra*, and *Welwitschia*, if available

Some Suggested Learning Goals

1. Understand the difference between the two types of leaves produced by pines and how pine leaves differ from those of other conifers.
2. Know the life cycle of a pine tree, and be able to indicate within the life cycle where events such as *meiosis, fusion of gametes, development of an embryo,* and *production of sperms* take place.
3. Understand the differences between male (pollen) and female (seed) pine cones.
4. Know the locations and functions of a pine *micropyle, integument, pollen chamber,* and *nucellus.*
5. Know the function of the *bladders* or *wings* on pine *pollen grains.*
6. Be able to distinguish a pine, a cycad, *Ginkgo, Gnetum, Ephedra,* and *Welwitschia* from one another (if they are available for examination).

Introduction

Although there are some exceptions among the algae and certain relatives of ferns, most of the plants studied thus far each produce just one kind of asexual spore. Beginning with the *gymnosperms,* however, the higher plants produce two distinct kinds, each within its own distinctive structure. Also, in Domains Archaea and Bacteria, Kingdom Protista, and the bryophytes of Kingdom Plantae, the primary means of dispersal is a *spore*. In gymnosperms and *angiosperms* (flowering plants), however, dispersal is primarily by means of *seeds,* which are considerably more complex and larger than spores. The word *gymnosperm* comes from two Greek words that mean "naked-seeded," in reference to the

fact that gymnosperm seeds are produced out in the open on cone scales, while the seeds of flowering plants are produced completely enclosed within fruits.

In contrast to the branching patterns of most broadleaf trees, the growth of most conifers is *excurrent* (i.e., the trunk of the tree does not divide unless the terminal bud is removed). Also, with the exception of *Ginkgo,* larch (*Larix*), and dawn redwood (*Metasequoia*), most gymnosperms have *evergreen* leaves. Unlike deciduous trees that seasonally lose all their leaves, most conifers lose a few leaves at a time. Nearly all conifers do have a complete change of leaves every 2 to 5 years. Bristlecone pines are an exception; they retain their leaves for about 30 years.

A. Conifers—Phylum Pinophyta

Carefully examine the youngest part of a stem of a pine branch. Note that two kinds of leaves are present. The most conspicuous leaves are needlelike and in *fascicles* (clusters) of 2, 3, or 5. With a sharp razor blade, cut one of the leaves, and examine the cut surface with your dissecting microscope. How many *vascular bundles* (veins) can you see? If you cut all the leaves in one fascicle and hold the cut remnants tightly together, do they form a complete cylinder? Are there small, inconspicuous, brownish scale leaves also present on the stem? How do pine leaves differ from those of the other conifer leaves provided? Are any of the other conifer leaves scalelike? Are the other conifer leaves arranged opposite one another or in a spiral on the stem?

Examine an open woody *seed cone* of a pine tree. Notice the paired, winged *seeds* at the base of each *cone scale*. The seeds in your particular cone may be missing; if so, the slight depressions in which the seeds were produced should be discernible. The seeds develop from *ovules,* in which a *megasporocyte* has undergone *meiosis,* producing *megaspores*. One megaspore develops into a *female* gametophyte that contains two or more *archegonia,* each with a large *egg* cell and a little nutritive tissue; the *nucellus,* above the *archegonia;* and other gametophyte tissue surrounding and below the archegonia. The female gametophyte is itself surrounded by a massive *integument* that later develops into the *seed coat* of a *seed*. Above the nucellus is a space called the *pollen chamber*. A somewhat tubular *micropyle* is located in the integument directly above the pollen chamber. A sticky fluid oozes through the micropyle to the outside, where it forms a *pollination drop*. A pollen grain may catch in the pollination drop and be slowly drawn into the pollen chamber as the fluid evaporates. Note that, unless your slide happens to be a specially selected median section, parts of one or both archegonia and/or the micropyle may have been cut off during manufacture. Once a pollen grain comes to rest above the archegonia, it may produce a *pollen tube* and two *sperms*. By the process of *fertilization,* one sperm unites with the egg, forming a *zygote,* which then develops into the *embryo* of a *seed.*

Pollen grains are developed from *microspores* produced when diploid *microsporocytes* in the *microsporangia* (sacs) at the base of the pollen cone scales undergo meiosis. Pollen cones are much smaller than their woody seed cone counterparts; they are usually produced in clusters at the tips of the lower branches of a tree. Mount a *small* amount of conifer pollen in a drop of water on a slide, and examine it with the aid of your compound microscope. Note the wings or bladders on each pollen grain. These give the pollen greater buoyancy in the wind.

When a seed germinates, the embryo within it develops into a new tree that constitutes the *sporophyte*.

B. Other Gymnosperms (Phyla Ginkgophyta, Cycadophyta, and Gnetophyta)

Examine the other representative gymnosperms on display. Note that *Ginkgo* has distinctive fan-shaped leaves with dichotomously forking veins. *Ginkgo* trees are *dioecious;* the male strobili (pollen cones) are produced only on male trees. The female trees, whose seeds have a fleshy covering with a nauseating odor, are produced singly instead of in cones. Unlike the sperms of pines, those of *Ginkgo* have flagella.

Cycads are slow-growing plants that have a trunk and large, palmlike leaves. Like *Ginkgo,* the species are dioecious; the sperms have thousands of spirally arranged flagella. Both the pollen and the seed cones of cycads can be very large, with some seed cones being more than a meter long and weighing over 200 kilograms at maturity.

There are about 70 known species of *gnetophytes* distributed among three genera. *Gnetum* species have broad leaves and are mostly vines and trees of the tropics. Half the species of gnetophytes are in the genus *Ephedra,* native to drier regions of southwestern North America. Most photosynthesis takes place in the stems of the shrubby *Ephedra* plants, which have tiny scalelike leaves. There is only one species of *Welwitschia,* a bizarre-looking plant confined to temperate desert regions of southwest Africa. *Welwitschia* produces two large straplike leaves from a somewhat cup-shaped "trunk." The leaves, which grow continuously from the base, flap in the wind and become split. Most of a *Welwitschia* plant's water supply comes from condensate of fog that rolls in from the ocean at night.

Drawings to Be Submitted

1. Draw a small branch of a pine, showing the needlelike leaves. Also draw the cone(s) present. Label FASCICLE, SCALE LEAVES, SEED CONE, and POLLEN CONE.
2. Draw a short part of a branchlet of one of the other conifers provided. Show not only the shapes of the leaves but also how they are arranged on the stem.
3. Diagram a section through a pine OVULE with the aid of the prepared slide provided. Label ARCHEGONIUM, NUCELLUS, POLLEN CHAMBER, MICROPYLE, and INTEGUMENT.
4. Label the following on the drawings of the life cycle of a pine tree provided: SPOROPHYTE, CLUSTER OF POLLEN CONES, MICROSPORANGIUM, MICROSPOROCYTE, MICROSPORES, POLLEN GRAIN, CONE SCALE, SEED CONE, MEGASPOROCYTE, MEGASPORE, DEVELOPING FEMALE GAMETOPHYTE, ARCHEGONIUM, EGGS, POLLEN CHAMBER, NUCELLUS, INTEGUMENT, SEED, EMBRYO SPOROPHYTE, and SEEDLING SPOROPHYTE.
5. Draw a pine POLLEN GRAIN, showing the WINGS or BLADDERS.
6. Draw a habit sketch of any one of the other gymnosperms available.

CONIFERS

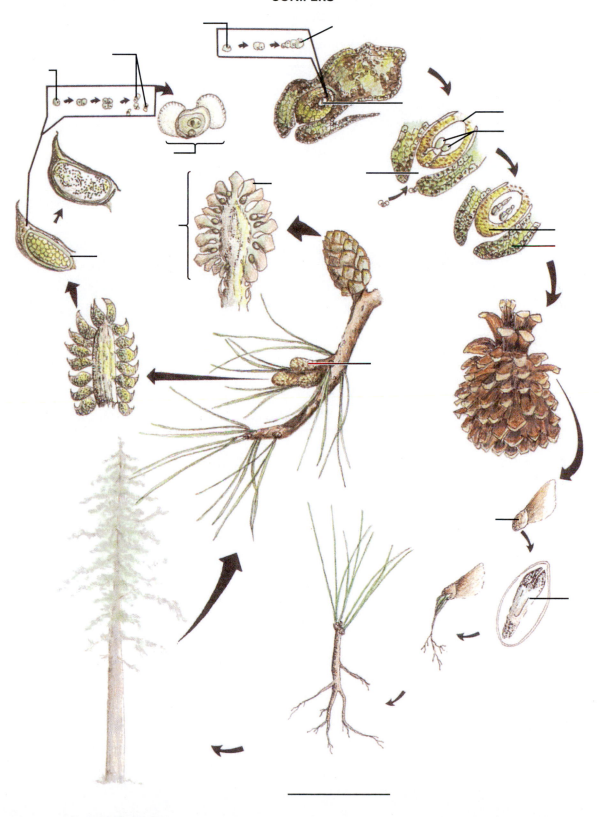

_____

_____

NAME _____

LAB SECTION NO. _____

DATE _____

OTHER GYMNOSPERMS

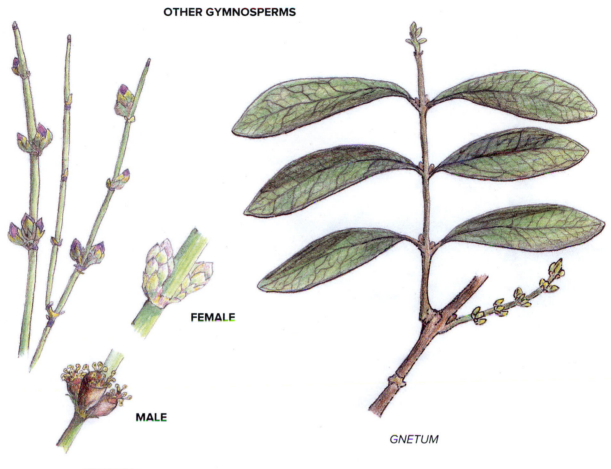

FEMALE

MALE

EPHEDRA

GNETUM

WELWITSCHIA

Review Questions 17

NAME _____

LAB SECTION NO. _____

DATE _____

1. What does the term *gymnosperm* mean, and in what sense does it apply to pine trees? _____

   _____

2. Apart from size differences, how can you distinguish a cone scale of a pine seed cone from that of a pine pollen cone?

   _____

3. Of what does the *female gametophyte* of a pine consist? What other tissues surround it? _____

   _____

4. What is the function of a *nucellus?* _____

5. What constitutes the *sporophyte* in a pine? _____

6. Where, specifically, are pine *pollen grains* produced? _____

7. What structure of a pine *ovule* develops into a seed coat? _____

8. What do the pollen grains of pine trees have that aid in their dispersal by the wind? _____

   _____

9. Could all the representatives of gymnosperms mentioned in this exercise be differentiated by their leaves alone? _____

   If not, why not, and if so, how? _____

   _____

Laboratory Preparation Quiz 17

NAME _____

LAB SECTION NO. _____

DATE _____

Kingdom Plantae: Gymnosperms

1. What is the difference between a *gymnosperm* and an *angiosperm?* _____

   _____

2. How many seeds are produced at the base of each pine seed cone scale? _____

3. What specific cells of pine undergo meiosis? _____

   What do these cells then become? _____

4. Through what passage is a *pollen grain* of pine drawn prior to its full development into a mature *male gametophyte?*

   _____

5. What is the space above the *nucellus* in a pine *ovule* called? _____

6. From what specific cell does the *embryo* of a *seed* develop? _____

7. Where on a pine tree are pollen cones usually produced? _____

8. From what structure does the *seed coat* of a pine seed develop? _____

9. Specifically, where are pine *sperms* produced? _____

10. Which of the gymnosperms discussed has two straplike leaves? _____

Kingdom Plantae: Angiosperms (Flowering Plants–Phylum Magnoliophyta)

Courtesy of James E. Bidlack

Materials

1. Large fresh flowers with superior ovaries
2. Prepared slides of *Lilium*—cross sections of ovary showing embryo sacs; *Lilium*—cross sections of mature anthers
3. Set of live flowers on display, including a composite, a grass, a primitive flower (e.g., buttercup), an advanced flower with an inferior ovary (e.g., orchid), and an inflorescence
4. Model of a flower

Some Suggested Learning Goals

1. Know the parts of a *complete flower,* and know the function of each part.
2. Understand the basic variations in *ovary* position and general structure of flowers.
3. Understand the difference between a *compound ovary* and a *simple ovary*.
4. Learn the life cycle of a flowering plant, and understand how an immature *ovule* becomes a *seed*.

Introduction

The life cycle of flowering plants exhibits the same alternation of gametophyte (*n*) and sporophyte (*2n*) generations seen in lower plants. However, the gametophyte phase is proportionately greatly reduced and is confined within certain tissues of the sporophyte, where it is entirely dependent on it. As in gymnosperms, two kinds of spores and gametophytes are produced, the smaller *microspores* giving rise to the male gametophytes, and the larger *megaspores* giving rise to the female gametophytes.

A. Structure of a Flower

With the aid of your dissecting microscope, examine one of the flowers provided. Note the small, leaflike *sepals,* comprising the *calyx,* and the usually colored or white *petals,* comprising the *corolla.* In some flowers, the sepals or petals may be united into an undivided calyx or corolla. How many sepals and petals does your flower have? *Monocot* flowers generally have their parts (sepals, petals, stamens, stigma divisions) in threes or multiples of three. *Dicot* flowers usually have their parts in fours or fives. Is your flower a monocot or a dicot?

Both the calyx and the corolla are attached to the *receptacle,* the slightly to distinctly expanded tip of the flower stalk (*peduncle* or *pedicel*). The male reproductive structures, or *stamens,* each consist of a slender stalk, the *filament,* and a pollen-bearing *anther.* In some flowers, the filaments may be fused together or to the petals; they usually surround the female reproductive structure, the *pistil.* The pistil consists of a *stigma* that may be knoblike, forked, feathery, or pointed; a necklike *style* that can be long and slender to short and stubby; and an *ovary,* which is usually swollen.

There are many variations of flower structure. In lilies, for example, the sepals may be of the same size and color as the petals. In the sunflower family, the "flower" is actually an *inflorescence,* composed of many tiny flowers arranged so that they resemble a single larger flower. Most of the tiny flowers have very small fused corollas and stamens, but those around the margin each may have a large flattened, petal-like extension of their corolla. In several families (e.g., the pumpkin family), the stamens and the pistil usually are in separate flowers; in other families, such as the buttercup family, there may be more than one pistil to a flower.

Remove the pistil from your flower, and note the swollen *ovary* at the base, the pollen-receiving *stigma* at the top, and the *style* connecting the two parts. Is your stigma divided in any way, or is it instead knoblike, or inconspicuous enough to be difficult to distinguish from the style? Now cut the ovary longitudinally with a razor blade. Observe the small whitish *ovules.* These eventually become *seeds* as the ovary matures into a *fruit.* Is your ovary divided into two or more segments known as *carpels?* Carpels represent individual pistils that have become united, and a pistil with two or more carpels is said to be *compound.* A *simple* pistil has only one carpel.

B. Development of the Male Gametophyte

Before young anthers mature, they usually contain four chambers in which *microsporocytes* undergo meiosis, producing quartets (sometimes called *tetrads*) of *microspores.* After the nucleus of each microspore has divided once by mitosis, the cells of each quartet separate, and their walls often become sculptured or ornamented. These bodies are now called *pollen grains.* As the anther matures, the wall between adjacent chambers usually breaks down, leaving just two *pollen sacs,* from which the pollen grains are released through slits or pores.

After *pollination* (which is nothing more than the transfer of pollen from an anther to a stigma and should not be confused with fertilization), a *pollen tube* may emerge from a pollen grain on the stigma, and by following a gradient of chemicals diffusing from the embryo sac, the pollen tube may grow down the style to the ovary and enter the ovule through an opening or passage called the *micropyle*. As the pollen tube grows, one of the two nuclei in the pollen grain, the *tube nucleus*, remains near the tip, while the other nucleus (*generative nucleus*) lags behind. Sometimes the generative nucleus divides in the pollen grain, forming two male gametes or *sperms*, but this particular mitotic division often takes place right in the pollen tube while it is growing. The germinated pollen grain, with its pollen tube containing two sperm nuclei, constitutes the mature *male gametophyte*.

Study the slide labeled "*Lilium:* mature anthers." Examine one of the sections of the anthers. Locate a pollen grain in which two nuclei are visible. Note the relatively thick, sculptured outer wall. The generative nucleus is more dense than the tube nucleus. How can you account for some of the pollen grains on your slide apparently having only one nucleus? (Hint: Remember that you are looking at very thin slices of tissue.)

C. Development of the Female Gametophyte

A *megasporocyte* in each *ovule* undergoes meiosis, producing four haploid *megaspores*. In most flowering plants, three of the megaspores degenerate. The remaining megaspore becomes larger as its nucleus undergoes three successive mitotic divisions. The three successive divisions result in eight nuclei. This large, eight-nucleate cell within the ovule constitutes the *female gametophyte*.

One of the eight nuclei, normally located toward the bottom of the female gametophyte, functions as the female gamete, or *egg;* the egg is flanked by two *synergid* nuclei. If the egg is damaged, either of the synergids can substitute as the egg. At the other end of the female gametophyte are three nonfunctional *antipodal* nuclei; the other two nuclei, called *central cell nuclei,* usually remain in the center of the female gametophyte where they sometimes fuse together.

Unless you have been provided with a specially selected slide, you probably will not see all eight of the female gametophyte nuclei. Most laboratories for introductory courses use *Lilium* (lily) ovary cross sections to show female gametophytes (Figs. 18.1A and 18.1B). Such a prepared slide usually contains several cross sections, and each section has parts of up to six gametophytes present. However, because of the way in which the sections are cut, complete gametophytes with all eight nuclei visible are seldom present. You should first locate the most complete gametophyte sections with the *low power* of your compound microscope, and then turn to high power to see details. Note that in a lily female gametophyte, four of the nuclei are considerably larger than the other four nuclei. This is because the lily gametophyte develops in a manner slightly different from that of most other flowers.

By the time it is mature, the diploid cells of the ovule surrounding it have developed into two layers, the *integuments,* which will later become the *seed coat* of a *seed*. There is normally a gap or passageway, the *micropyle,* formed between the integuments in the vicinity of the egg. This micropyle will later allow access to the female gametophyte by a tube from a pollen grain.

Examine the slide of a *Lilium* mature embryo sac, and locate the ovary, which contains three chambers. Within each of these chambers, you may see two ovules. Locate one ovule, and look for the two integuments and a small opening known as the *micropyle* between them. If you have chosen a good

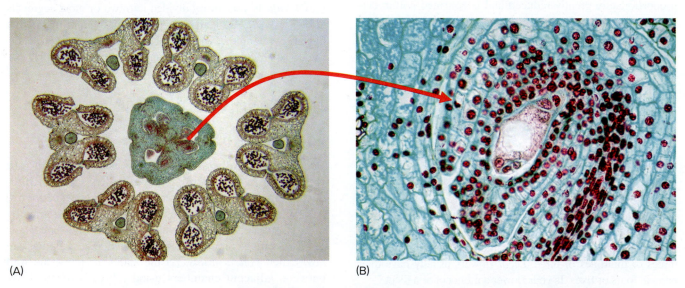

(A) (B)

FIGURE 18.1   (*A*) CROSS SECTION OF A LILY FLOWER SHOWING SIX ANTHERS SURROUNDING A CENTRAL OVARY, WHICH IN TURN, IS COMPOSED OF SIX OVULES. ×10. (*B*) CROSS SECTION OF ONE LILY OVULE, SHOWING A MATURE EMBRYO SAC AND SEVERAL OF THE FEMALE GAMETOPHYTE NUCLEI. ×400. *Courtesy of James E. Bidlack*

cross section, you should be able to see several of the female gametophyte nuclei. Can you determine which of these nuclei are antipodals, synergids, and central cell nuclei? Which nucleus should fuse with the sperm to form a zygote?

D. Development of Seeds and Fruits

After pollination and growth of the pollen tube has occurred, the contents of the pollen tube are discharged into the embryo sac. The *double fertilization* that follows involves the union of one sperm with the egg, forming a *zygote,* and the union of the other sperm with the polar nuclei, forming the *endosperm nucleus.* Because the endosperm nucleus is the product of three haploid (*n*) cells all fusing together, it ends up *triploid* (i.e., it has 3*n* chromosomes)—a situation unique to the flowering plants. The endosperm nucleus usually divides repeatedly, producing 3*n* endosperm tissue, which functions in food storage. The endosperm may become an extensive part of the seed, or it may disappear soon after it is formed. When the endosperm disappears early, part of the *embryo,* which develops by repeated divisions of the zygote, may take over the food storage function. The outer layers of the ovule (integuments) harden into a *seed coat,* which forms the outer covering of a seed. While the seeds are maturing, the ovary undergoes transformation into a *fruit.*

Drawings to Be Submitted

1. Label all the parts of a complete flower. Include PEDUNCLE (or PEDICEL), RECEPTACLE, SEPALS (CALYX), PETALS (COROLLA), STAMEN(S) with ANTHER(S) and FILAMENT(S), PISTIL, STIGMA, STYLE, OVARY, and OVULE.
2. Draw a portion of a cross section of a MATURE ANTHER from the prepared slide provided. Label POLLEN GRAIN(S), TUBE NUCLEUS, and GENERATIVE NUCLEUS.
3. Draw a section through an OVULE, using the prepared slide provided. Show the EMBRYO SAC, with its ANTIPODALS, SYNERGIDS, and EGG, and the surrounding tissues.
4. Label the following on the drawings of the life cycle of a flowering plant provided: SPOROPHYTE, FLOWER, STAMEN, ANTHER, FILAMENT, PISTIL, STIGMA, STYLE, OVARY, MICROSPOROCYTE, MICROSPORES, MEGASPOROCYTE, MEGASPORE, FEMALE GAMETOPHYTE, DIVISIONS OF MEGASPORE NUCLEUS, OVARY (showing embryo sac), ANTIPODALS, SYNERGIDS, INTEGUMENTS, POLLEN GRAINS, (STIGMA, STYLE, and OVARY) with POLLEN TUBE, EGG, ZYGOTE, EMBRYO, FRUIT, and SEED. Also indicate where MEIOSIS occurs and where FERTILIZATION takes place.

Exercise 18

NAME _____

LAB SECTION NO. _____

DATE _____

FLOWERING PLANTS

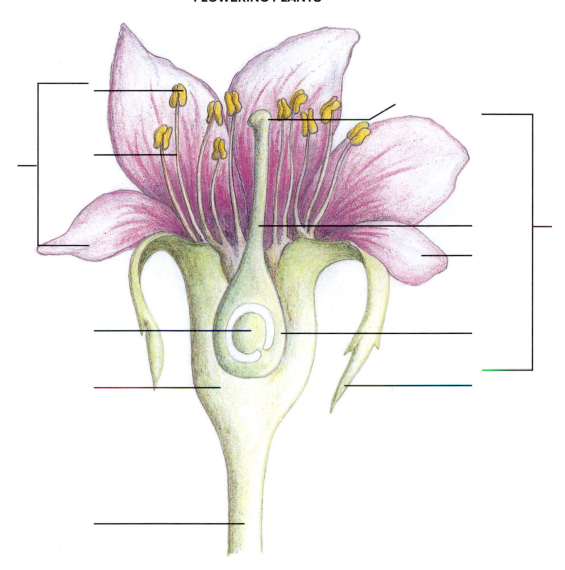

FLOWERING PLANTS

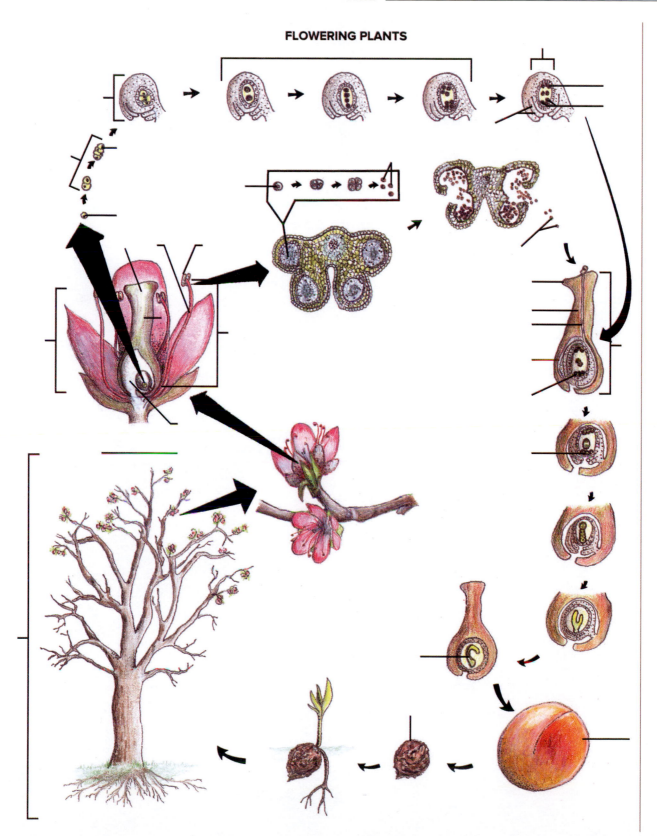

Review Questions 18

NAME _____

LAB SECTION NO. _____

DATE _____

1. What is the part of the flower to which *petals, sepals,* and *stamens* are attached? _____
   _____

2. What is the part of the stamen to which an *anther* usually is attached? _____

3. Which part of a flower receives *pollen?* _____

4. How many functional cells are usually produced when a *microsporocyte* undergoes meiosis? _____

   What are these cells called and what happens to them? _____

5. How many *nuclei* does a typical mature pollen grain have before pollination? _____

   How many nuclei does a mature male gametophyte have just before fertilization? _____
   _____

6. How does a *pollen tube* enter an *embryo sac?* _____
   _____

7. What is the difference between *pollination* and *fertilization?* _____

8. What happens to *antipodals* after fertilization has occurred? _____
   _____

9. How is an *endosperm nucleus* formed? _____

   What becomes of an endosperm nucleus after it is formed? _____

10. What structures become a *seed coat?* _____

Laboratory Preparation Quiz 18

NAME _____

LAB SECTION NO. _____

DATE _____

Kingdom Plantae: Angiosperms (Flowering Plants–Phylum Magnoliophyta)

1. What do we call a group of flowers sharing a common main stalk? _____

_____

2. What are the *whitish objects* that eventually become *seeds* inside an ovary? _____

3. What is the name of the nuclei that flank the *egg* in a female gametophyte? _____

4. What is a *micropyle?* _____

5. How many nuclei does a newly formed *microspore* have? _____

6. When a *generative nucleus* divides, what are the resulting cells or nuclei called? _____

_____

7. With which nuclei does a sperm unite, forming an *endosperm nucleus?* _____

8. Which cell undergoes meiosis, producing *megaspores?* _____

9. Specifically where in a flower is the *female gametophyte* located? _____

10. Where does *fertilization* occur in a flower? _____

Fruits, Spices, and Survival Plants

19

Materials

1. As complete as possible representation of both dry and fleshy fruit types
2. Demonstration of as many spices as is feasible
3. Display of survival plants (either herbarium specimens or live material) indicating how each plant is used
4. Display of poisonous plants (either herbarium specimens or live material) with notes on toxicity

Some Suggested Learning Goals

1. Know how the various fruit types are distinguished from one another.
2. Know at least one or two representatives of each fruit type.
3. Learn which plant parts are used for specific purposes.
4. Know several representative survival plants, how they are used, and how the toxic plants poison humans.

Introduction

In this exercise, you will be asked to classify a number of fruits and to know examples of each fruit type. You will also be asked to take notes on what plant parts are used to prepare the various spices on display, know how the survival plants are used, and be able to give examples of common poisonous plants.

A. Fruits

Fruits may be defined as mature ovaries, usually containing seeds. At maturity, the ovaries swell and become *fleshy* ("juicy"), or the ovary wall shrivels and becomes *dry*. The separation between dry and fleshy fruits is occasionally indistinct, but most fruits are relatively easy to categorize in this regard. In dry fruits, either the seeds are released when the fruit splits at maturity, or the fruit wall layers adhere to the seeds within and remain until the seed germinates. In some instances, these fruit wall layers form wings that aid in dispersal; such wings should not be confused with the wings of conifer seeds, which are merely extensions of the seed coat.

A mature fruit, such as a peach, has three identifiable regions in addition to the seed(s). The outermost region ("skin") is referred to as the *exocarp*. The innermost region (the pit in a peach) is called the *endocarp;* it usually surrounds the seed(s). The flesh between the exocarp and the endocarp is the *mesocarp*. In dry fruits, in particular, two or all three of the layers may be fused together and, collectively, are referred to as the *pericarp*. Some fruits, such as apples and strawberries, consist of other flower parts in addition to the ovary or ovaries. For example, the calyx itself may become fleshy and comprise more extensive fruit tissue than the ovary that it surrounds. A fruit that consists of more than just the ovary is said to have *accessory* tissue or is called an *accessory* fruit.

The fruit key that follows gives you many of the primary distinctions among the various kinds of fruits, but it is not completely practical unless you have more information about the flowers from which the fruits are derived. Your instructor will help you with whatever additional information is needed to arrive at the correct botanical classification for the displayed fruits.

Key to Common Fruits

1a. Fruits fleshy.
    2a. Fruits simple (i.e., derived from a flower with a single pistil).
        3a. Fruits with a single seed enclosed in a hard pit (cherry, peach) . . . . DRUPES
        3b. Fruits with more than one seed, the seeds not enclosed in a hard pit (only one seed develops in an avocado, and no seeds develop in the common banana).
            4a. Fruits with thin or leathery skin or the outer part of the fruit forming a rind; endocarp not leathery or papery (banana, tomato) . . . . BERRIES
            (Berries with a thin skin are referred to as *TRUE BERRIES;* berries with a leathery skin containing oils are referred to as *HESPERIDIUMS;* berries with a rind are referred to as *PEPOS.*)
            4b. Fruits with leathery or papery endocarps (apple, pear) . . . . POMES
    2b. Fruits derived from more than one pistil.
        5a. Fruits derived from a single flower having several to many pistils (blackberry, raspberry) . . . . AGGREGATE FRUITS

5b. Fruits derived from several to many separate flowers in an inflorescence,
the fruits coalescing to varying degrees to form a single "fruit" at maturity
(mulberry, pineapple) . . . . MULTIPLE FRUITS
1b. Fruits dry at maturity.
  6a. Fruits not splitting at maturity.
    7a. Fruits with a wing (elm, maple) . . . . SAMARAS
    7b. Fruits without a wing.
      8a. Fruits with a hard shell surrounding the seed (acorn, chestnut) . . . . NUTS
      8b. Fruits without a hard shell.
        9a. Fruit wall fused to the seed coat (rye, wheat) . . . . GRAINS (CARYOPSES)
        9b. Fruit wall with seed loosely attached (strawberry, sunflower). ACHENES
  6b. Fruits splitting in various ways at maturity.
    10a. Fruits splitting along or between carpel lines or forming a cap that comes off or a row of pores
near the top (lily, poppy). . . . . CAPSULES
    10b. Fruits splitting lengthwise along the edges.
      11a. Fruits leaving a central partition to which the seeds are attached
(cabbage, mustard) . . . . SILIQUES or SILICLES
      11b. Fruits not leaving a central partition.
        12a. Fruits splitting along one edge only (larkspur, milkweed) . . . . FOLLICLES
        12b. Fruits splitting along both edges (bean, pea) . . . . LEGUMES

To gain a better understanding of fruit structure, your instructor may provide you with one or more fruits for dissection. Some common fruits found at the grocery store include peaches, tomatoes, oranges, and beans. Their dissection and identification are described as follows.

1. Peach: Cut a peach in half longitudinally, and observe the skin (*exocarp*) and fleshy *mesocarp*. The *endocarp* forms a darkened pit that protects and surrounds the seed. The groove that encircles the fruit represents a suture between fused carpels. This peach, which has a single seed enclosed in a hard pit, is a typical *drupe*.

2. Tomato: Cut a tomato in half, and note the three identifiable regions within the fruit. The thin outer skin is the exocarp. Inside the exocarp is the fleshy mesocarp, and closest to the seeds is the endocarp. Much of the fleshy portion is composed of a large central *placenta,* the region of ovary to which the seeds are attached. The seeds are numerous and found within individual segments called *locules*. The tomato is classified as a *berry* because it is fleshy throughout and contains more than one seed.

3. Orange: Cut an orange in half to expose the inside of the fruit. The colored skin (exocarp) is made up of epidermis and some compact parenchyma cells, which contain oil glands and crystals. The oil glands should be visible with the aid of a dissecting microscope, in a ring around the periphery just inside the epidermis. The white portion of the fruit wall is the mesocarp. Partitions between individual segments of the orange (derived from carpels) constitute the endocarp. The carpels are packed with engorged juice sacs, which form as outgrowths of the endocarp. Because it has a leathery skin containing oils, the orange is a special kind of berry called a *hesperidium.*

4. Bean: Split a bean pod in half along its margin so you can see the fruit wall and the seeds. Exocarp, mesocarp, and endocarp are fused together but discernible by careful observation. The placenta can be identified between individual seeds and the endocarp. In beans and other dehiscent fruits that open by splitting, sclerenchyma cells within the three fruit layers are oriented in different planes. As the fruit dries, the cells of each layer shrink along their longitudinal axes, creating shear forces that cause the pericarp to twist and finally split along lines of weakness. The bean is defined as a *legume* because it is dry at maturity and splits lengthwise along both edges.

B. Spices

Spices are derived from a wide variety of plants and plant parts, although some plant families seem to produce more of them than others. Make a list of the spices and flavoring materials on display, and indicate which part of the plant (e.g., leaves, flowers) is used as a spice.

C. Survival Plants

The number of plants grown for food, medicine, and other purposes relating to survival is very large, and past cultures made even more extensive use of plants than we do today. It is possible here to examine only a tiny fraction of the local representatives. In addition to making a list of the plants on display, make notes as to what parts of the plants have been or can be used by humans, and indicate the specific uses.

Review Questions 19

NAME _____

LAB SECTION NO. _____

DATE _____

1. What distinguishes a *hesperidium* from a *pepo?* _____

   _____

2. Which of the fruit types is derived from more than one pistil? _____

   _____

3. How do you tell a *grain (caryopsis)* from an *achene?* _____

   _____

4. If you were to cut an apple in half, you would notice that the *endocarp* around the seeds is somewhat papery. How would

   you classify it as to fruit type? _____

5. *Drupes* and *nuts* both have a single seed. What distinguishes them from one another? _____

   _____

6. Black raspberries and mulberries look quite a bit alike, but raspberries are *aggregate fruits* while mulberries are *multiple*

   fruits. What is the difference? _____

   _____

7. Choose one of the poisonous plants, and tell which part or parts is (are) poisonous. _____

8. When you use oregano as a spice, what part of the plant is involved? _____

9. Name two spices that are derived from flowers or flower buds. _____

   _____

10. Choose one of the survival plants, and tell how it is used. _____

    _____

Laboratory Preparation
Quiz 19

NAME _____

LAB SECTION NO. _____

DATE _____

Fruits, Spices, and Survival Plants

1. Which fleshy fruits have a single seed enclosed in a hard pit? _____

2. Strawberry flowers have numerous pistils on a common receptacle. What fruit type does that make them?

   _____

3. How many seeds does a typical *berry* have? _____

4. What distinguishes a *hesperidium* from a *true berry?* _____

   _____

5. Do both *aggregate* and *multiple fruits* come from more than one pistil? Explain. _____

   _____

6. Give a common example of a fruit in which the seeds do not develop. _____

7. Which dry, splitting fruit has a central partition to which the seeds may be attached? _____

8. What type of dry, nonsplitting fruit has a wing at maturity? _____

9. What type of dry, splitting fruit splits only along one edge? _____

10. How do you tell a *grain* (caryopsis) from an *achene?* _____

   _____

Ecology

Materials

1. Two long, metal measuring tapes or large balls of string
2. Marking pen

Some Suggested Learning Goals

1. Explain how plants can be affected by the environment.
2. Learn the anatomical modifications of *hydrophytes* and *xerophytes*.
3. Know the meaning of *succession,* and be aware of its various forms.
4. Understand what a *transect* is and how it can be used for analysis of vegetation.

Introduction

Ecology is a vast field of study involving many aspects, particularly the relationships among organisms and their environment. It is a complex topic that integrates virtually all levels of biological organization and how they respond to each other and to nonliving components of their surroundings. Through investigations of ecology, biologists have a better understanding of how organisms interact with each other and how changes in their surroundings may affect the sensitive balance in various ecosystems. Some types of ecological studies involve sophisticated models and statistical analyses that are beyond the scope of this course. However, some aspects of ecology can be introduced by observation of organisms in their natural habitats and collection of data to indicate how they respond to certain environments. In this exercise, you will learn about adaptations of plants to specific habitats and collect data to learn about diversity, population density, and growth dynamics in different situations.

Adaptations of Plants to Specific Habitats

Plants that grow in or near water or that float on water are called *hydrophytes*. Hydrophytes tend to have significantly less xylem than *mesophytes* (plants that grow primarily on land and have a moderate supply of water). Consequently hydrophytes are seldom rigid out of water, the water itself providing some of the needed support. Hydrophytes often have extensive interconnecting air spaces within their tissues, and those with floating leaves tend to have the stomatal pores confined to the upper epidermis. *Xerophytes*

are found in deserts and other areas where water may not be available for months at a time. Xerophytes have several modifications that reduce their water loss. Such modifications include sunken stomata, leaves modified as spines, thick wax coverings, dense hairs, and water-storage cells. *Halophytes* are plants that normally grow in soils with high salt content. Halophytes may have some characteristics of xerophytes because the salt interferes with osmosis so that the amount of water actually absorbed is no more extensive than the amount absorbed by xerophytes.

A. Succession

Depending on the location of your institution, it may be possible to take a field trip to a *hydrosere* or a *xerosere*. A hydrosere is a series of stages of *succession* in a pond, lake, or other body of water. Succession in a wet habitat is the gradual directional shift, over time, from a wet environment to a *mesic* (in-between or intermediate) environment, brought about by the activities and interactions of the organisms that constitute the living part of the habitat. Succession also takes place in environments where there is little water present at the start. A *xerosere* involves a series of different stages of succession that begin with bare rock or cooled lava. Other successions (Fig. 20.1) that could be studied might include a field that has been abandoned after it has been overgrazed; a parking lot that has been abandoned; a burned forest or grassland; a roadcut; or an abandoned swimming pool.

B. Transects

To determine, statistically, the nature of vegetation or its *biomass* (total weight) at a given site or to find out if vegetation varies statistically on, for example, a north- versus a south-facing slope, biologists might run a *transect* through the area. A transect is essentially a line or area between two parallel lines, with the length varying according to what is to be analyzed. The distance between two parallel lines can also vary, although a popular transect width is 1 meter. Then the transect is divided into lengths of 1 meter or 1 foot (or whatever unit of measure suits the subject). For single-line transects, all plants touching the line are counted and analyzed, or random samplings of a given unit of length are made along the transect. Sometimes soil is analyzed in each unit of the transect, and temperature and light readings may be kept. Over more extended periods, records

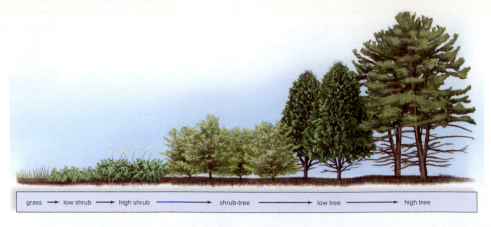

grass → low shrub → high shrub ──────→ shrub-tree ────→ low tree ────→ high tree

FIGURE 20.1 SECONDARY SUCCESSION IN A FOREST THAT OCCURRED IN A LARGE CONIFER PLANTATION IN CENTRAL NEW YORK STATE.

of precipitation may also be accumulated. In the wider transects, the area may be divided into squares, and all the plants in the squares may be counted and categorized. If the transect is quite long, every second or third square may be analyzed.

Assignments

1. On your succession field trip, make careful notes on the kinds of plants and other organisms encountered in either a hydrosere or a xerosere (or other successional area), and take small samples, being very careful not to disrupt the natural ecology. Record the numbers of different hydrophytes and xerophytes you find (if your instructor is able to identify them for you, take notes on this information also). Note the relative locations of the various kinds of plants, and record the information on a diagram. Make stem sections of any plants you collect, and examine them with a microscope. Take notes on any unusual features, such as large air spaces, dense hairs, waxy surfaces, etc., and record your observations. Then prepare a formal report of your activities, to be turned in within 2 weeks.

2. Run a line transect (by unrolling a measuring tape or a ball of string in a straight line) at the location to which your instructor directs you. Mark off lengths of 1 meter along the entire transect. On every other meter length, record the plants that are touching the tape or string. Because you may not know the names of specific plants, at least categorize them as grasses, other herbaceous plants, shrubs, and trees. After you have completed your recording, total the numbers in each category, and indicate what percentage of the whole they represent. Then prepare a formal report of your experience, putting your statistics in the form of a table. Turn in the report at the next laboratory session.

3. Your instructor may encourage you to visit one or more websites to learn about plant succession. Some good websites on this topic can be found by searching the word "succession," at http://www.merlot.org. Some of the most interesting websites that provide pictures and up-to-date research projects on succession have been devoted to studies after volcanic eruptions at Mount St. Helens. Studies of plant succession at Mount St. Helens can be found at the following links:

http://faculty.washington.edu/moral/mount.html
http://volcano.oregonstate.edu/Mount_St_Helens
http://www.fs.fed.us/gpnf/mshnvm/
http://www.fsl.orst.edu/msh

Review Questions 20

NAME _____

LAB SECTION NO. _____

DATE _____

1. What is a *hydrosere?* _____

_____

2. How would you describe a *mesic* environment? _____

_____

3. Which type of *succession* begins with bare rock? _____

4. List three modifications of *hydrophytes.* _____

_____

5. In what type of environment does a *halophyte* grow? _____

6. List three modifications of *xerophytes.* _____

_____

7. Why do some halophytes resemble some xerophytes in appearance? _____

_____

8. What is *biomass?* _____

9. What is a *transect?* _____

10. How is a transect used? _____

Laboratory Preparation Quiz 20

NAME _____

LAB SECTION NO. _____

DATE _____

Ecology

1. What do we call plants that have adaptations for growing in or near water? _____

2. Would having a waxy surface be an adaptation of a *hydrophyte*, a *mesophyte*, or a *xerophyte*? Explain. _____

   _____

3. How is a *succession* defined? _____

   _____

4. What term do we apply to a plant adapted to growing in the presence of higher than normal amounts of salt? _____

   _____

5. Which type of succession begins with a body of water? _____

6. What does *mesic* mean? _____

7. Which tissue of mesophytes is generally less abundant in hydrophytes? _____

   _____

8. What term is applied to the total weight of organisms in a given area? _____

9. What is a *transect?* _____

   _____

10. Is there any particular limit to the width of a transect? _____

Genetics

Materials

1. F$_2$ generation ears of corn produced from parents that were homozygous red, starchy, and homozygous white, sweet
2. Coins

Some Suggested Learning Goals

1. Know the meaning of basic genetic terminology such as *F$_1$*, *F$_2$*, *genotype, phenotype, homozygous, heterozygous, dominant, recessive, gene,* and *hybrid*.
2. Understand how to calculate probability with coins.
3. Be able to diagram both *monohybrid* and *dihybrid crosses*.
4. Be able to work and solve basic genetics problems.

Introduction

The following simplified definitions of terms may help to refresh your memory for this exercise:

allele—One of a pair of genes at the identical location (*locus*) on a pair of homologous chromosomes.

dominant—One member of a pair of genes is said to be dominant over the other when it completely masks or suppresses the expression of the alternate characteristic.

F$_1$ generation—The offspring of a cross between two parents.

F$_2$ generation—The offspring of crosses between F$_1$ individuals.

gene—A unit of DNA, carried on a chromosome.

genotype—The genetic constitution of an individual.

heterozygous—Having contrasting members of a given pair of alleles in an individual.

homologous chromosomes—Pairs of structurally identical chromosomes that associate together during prophase I of meiosis; each member of a pair is derived from a different parent.

homozygous—Having identical members of a given pair of alleles in an individual.

hybrid—The offspring of a cross between two unlike individuals.

parental generation—The parents producing the gametes involved in the production of an F$_1$ generation.

phenotype—The physical appearance of an individual.

recessive—One member of a pair of alleles is said to be recessive when its expression is completely masked by its dominant gene.

A. Probability

The Austrian monk Gregor Mendel, whose careful studies in the 19th century laid the groundwork for the science of genetics, experimented with a number of garden plants. In peas, he found that tallness (*T*) is dominant, while dwarfness (*t*) is recessive. If two heterozygous tall pea plants are crossed, what proportion of the offspring should be tall? Should it be possible to deviate from this expected ratio? An answer to this can be obtained by repeatedly flipping coins (to represent the gene pairs) and keeping track of the results.

Flip two coins simultaneously. Assume that heads represent the dominant gene (*T*) and tails the recessive gene (*t*) and that each coin represents one of the heterozygous parents. When a coin is flipped, it will land as either a *T* gamete (head) or a *t* gamete (tail). The equivalent of mating is achieved by flipping both coins simultaneously, and the faces showing would be equivalent to the genotype of an offspring. Continue to flip coins until you have made a total of 100 flips, and enter your results on the first chart on the following page. To calculate the ratios, use the following formula:

$$\frac{4 \times \text{number of flips of a given genotype}}{\text{total number of flips}} = \text{ratio}$$

B. Dihybrid Cross in Corn

In corn, the kernel colors red and white are controlled by a single pair of genes located on a particular pair of chromosomes; the expression of starch (smoothness) and sweetness (being wrinkled) is controlled by another pair of genes located on a different pair of chromosomes. The ears of corn for this exercise are from the F$_2$ generation of a cross originally between parents that are homozygous red, starchy, and parents that are homozygous white, sweet. *Without removing the kernels from the ear,* count and record the various phenotypes present on the second chart on the following page.

	Your 100 Flip Totals			Class Totals	
Genotype	Expected	Observed	Ratio	Observed	Ratio
TT	25				
Tt	50				
tt	25				
Totals	100	100	4		4

	Observed		Ratio	
Phenotype	Your Ear	Class	Your Ear	Class
Red, smooth				
Red, wrinkled				
White, smooth				
White, wrinkled				
Totals			16	16

1. Calculate the ratios based on a total of 16. To do so, use this formula:

$$\frac{16 \times \text{number of kernels of a given phenotype}}{\text{total number of kernels in the ear}} = \text{ratio}$$

2. Diagram the genotypes of the F_2 in the following boxes:

Gametes				

a. What was the phenotype of the F_1 generation? _____

b. What phenotype(s) was (were) found in the F_2 generation? _____

_____

3. What phenotypes, and in what ratio, would we expect from the following cross?

$$SSWw \times ssWw$$

_____

_____

_____

_____

Additional Genetics Problems

1. Radishes may be oval, long, or round in shape. Crosses between oval and round produced 181 round and 177 oval; crosses between oval and long gave 202 oval and 206 long. Crosses between round and long gave 545 oval. Crosses between oval and oval gave 118 long, 240 oval, and 121 round. What type of inheritance is involved?

2. In summer squashes, white fruit (W) is dominant over colored fruit (w), and dish-shaped fruit (D) is dominant over globe-shaped fruit (d). How many different genotypes may squash plants have with regard to these two pairs of characteristics?

3. One variety of tomato has fine hairs on its skin. When a typical smooth-skinned tomato was crossed with a hairy tomato, all of the offspring had smooth skin. Crosses between the F_1 smooth-skinned tomatoes, however, produced 221 hairy tomatoes and 656 smooth-skinned tomatoes. How are the skin types inherited?

4. Assume that in cornflowers, blue flower color is dominant and white is recessive. Further assume that double flowers are dominant, and single flowers are recessive. If a heterozygous blue, double-flowered variety is crossed with a homozygous white, single-flowered variety, what types of offspring would be produced, and in what proportions?

5. How many different kinds of genotypes could be produced by the cross *HHBb* × *Hhbb?*

6. Garden plants produce many shades of color in their flowers. Two well-known plants, peas and snapdragons, produce both red and white flowers. When a homozygous red pea plant is crossed with a homozygous white pea plant, the F_1 generation has only red flowers, but the F_2 generation produces flowers in a ratio of 3 red to 1 white. A similar cross between snapdragons, however, produces an F_1 generation that is all pink and an F_2 generation of 1 red to 2 pink to 1 white. How may the difference be explained?

7. When a tomato plant with fruits that are solid red when ripe is crossed with a variety whose fruits have green at the top when ripe, all of the F_1 generation produce fruits with green at the top. What are the genotypes of the parents?

8. If a homozygous recessive parent is crossed with an F_1 offspring with the genotype *HhTt,* what genotypes will be produced? Give the ratio.

9. Assume that long stems and the production of fragrance in certain roses are dominant traits and that short stems and lack of fragrance are recessive. If two long-stemmed, fragrant roses—one homozygous and the other heterozygous—are crossed, what phenotypes and genotypes would you expect in the F_1 generation?

10. Diagram the genotypes from the F_2 generation of *MMGG* $\times$ *mmGg*:

Review Questions 21

NAME _____

LAB SECTION NO. _____

DATE _____

1. What term is applied to the *physical appearance* of an individual? _____

2. What is an *allele?* _____

3. What is a *recessive allele?* _____

4. How is *genotype* defined? _____

5. With what plants did Gregor Mendel do his most famous work? _____

6. How is the equivalent of mating achieved with coins? _____

_____

7. How are ratios calculated from the results of your coin flipping? _____

_____

8. What are the *dominant* traits in the corn dihybrid cross? _____

_____

9. Which fruits and vegetables are used for the genetics problems? _____

_____

10. Diagram the *genotypes* of the F_2 generation of the following cross:

$$KKPP \times kkpp$$

Laboratory Preparation Quiz 21

NAME _____

LAB SECTION NO. _____

DATE _____

Genetics

1. What do we call the offspring of a cross between two unlike individuals? _____

2. What is a *phenotype?* _____

3. Who was the Austrian monk who laid the groundwork for the science of genetics? _____

    _____

4. In this exercise, which side of a coin represents a *recessive allele?* _____

5. What is calculated with this formula? $\dfrac{4 \times \text{number of flips of a given genotype}}{\text{total number of flips}}$ _____

    _____

6. In flipping your coins, how many times out of 100 would you expect them to land with the *Tt* combination? _____

7. In the corn dihybrid cross exercise, what traits are *recessive?* _____

    _____

8. How many different *kinds* of genotypes would you expect in the corn dihybrid cross? _____

9. What is meant by the term *heterozygous?* _____

    _____

10. How many letters does a gamete of the corn dihybrid cross have? _____

Keys to Trees, Shrubs, and Vines

A Leaf Key to Some Trees, Shrubs, and Vines Commonly Found on College and University Campuses

Glossary

Axil—the *angle* formed between the stem and the leaf stalk (petiole)

Axillary—located in an axil

Bipinnate—leaf having both primary and secondary divisions of the blade

Bundle scar—a tiny scar within a leaf scar, left by xylem and phloem

Compound leaf—one that is subdivided into leaflets

Gland—a tiny bump or depression that may be functionless or may secrete various substances

Leaflet—an individual division of the blade of a compound leaf

Lenticel—a small structure in the bark through which gas exchange occurs

Lobe—a partial division of a leaf or leaflet blade; may be rounded, resembling a large tooth

Margin—the boundary or edge of a leaf blade

Midrib—the conspicuous central vein of a leaf or leaflet

Node—a region on a stem at which a leaf or leaves arises (arise)

Palmate—lobed or divided so that the main veins all converge at the junction of the leaf blade base and the petiole

Petiole—the stalk of the leaf

Pinnate—having secondary veins or leaflets diverging from the midrib, or rachis

Rachis—the equivalent of a midrib in a pinnately compound leaf

Simple leaf—one in which the blade is not divided into leaflets

Stipules—appendages that occur in pairs toward the base of a petiole; may fall off while the leaf is young, may be absent, or may be modified as spines

Tendril—a modification of a leaf, leaflet, or stem that aids in climbing

Whorled—having more than two leaves at a node

Note: When using leaf keys, examine leaves from several areas of the plant, since the leaves may vary in size and shape. Leaves from "sucker" shoots, which often arise around the base of the plant, are particularly variable and should be avoided. If keying a plant that is developing new growth, try to examine the most mature twigs, as internodes may not elongate to their normal lengths until some time after emerging from the buds.

1a. Leaves alternate or in a spiral.
  2a. Leaves simple.
    3a. Leaves fan-shaped, with numerous fine, forking veins; notch present in middle of broad end .... *GINKGO*
                                                                  Ginkgo biloba

    3b. Leaves not as above.
      4a. Leaves entire (i.e., without distinct teeth or lobes).
        5a. Petiole with bladelike wings and jointed below base of blade;
            foliage with aromatic oils .... *CITRUS* (*ORANGE, LEMON*, etc.)
                                         Citrus spp.

      5b. Petiole not winged or jointed.
        6a. Fine line (stipule scar) encircling twig at each node.
          7a. Leaves tough and leathery; lower surface of young leaves often with
             rust-colored hairs ....*SOUTHERN MAGNOLIA*
                                      Magnolia grandiflora

          7b. Leaves not tough and leathery; rust-colored hairs absent ....*JAPANESE MAGNOLIA*
                                      Magnolia soulangeana

        6b. Fine line (stipule scar) not encircling twig at each node.
          8a. Leaves circular to heart-shaped in outline .... *REDBUD*
                                      Cercis spp.

        8b. Leaves not circular to heart-shaped in outline.
          9a. Leaves with distinct medicinal or aromatic odor when crushed.
            10a. Leaves with single main vein from the base.
              11a. Leaves with strong odor of bay; yellowish green and somewhat
                 shiny ....*CALIFORNIA BAY*
                                      Umbellularia californica

            11b. Leaves not as above.
              12a. Leaves curling slightly toward the edges, with fine pale
                 yellow margins ....*CAMPHOR TREE*
                                      Cinnamomum camphora

            12b. Leaves flattened; with odor of eucalyptus oil ....*EUCALYPTUS*
                                        Eucalyptus spp.

          10b. Leaves with three main veins from near the base (some leaves may be lobed)...........*SASSAFRAS*
                                      Sassafras spp.

        9b. Leaf odors various when crushed but not distinctly medicinal or aromatic.
          13a. Virtually all leaves alternate.
            14a. Leaves tending to curl under slightly at the margins; leathery in texture .... *PITTOSPORUM*
                                      Pittosporum spp.

            14b. Leaves not in all respects as above.
              15a. Single large vascular bundle present in the petiole .... *PERSIMMON*
                                      Diospyros spp.

             15b. More than one petiole vascular bundles.
              16a. Leaves evergreen (persistent).
                17a. Spine present at some of the nodes; some leaves shallowly
                   notched at the tip or with minute hair point.. *PYRACANTHA*
                                        Pyracantha spp.

17b. Spines absent; leaves not as in 17a.
 18a. Leaves mostly less than 5 cm (2 in.) long; fruit an acorn...................... *LIVE OAK*
                           Quercus spp.

 18b. Leaves mostly more than 5 cm (2 in.) long;
   fruit a capsule... RHODODENDRON
                        Rhododendron spp.

16b. Leaves deciduous.
 19a. Spines or thorns usually present.
  20a. Trees; leaves and twigs with milky sap .... *OSAGE ORANGE*
                   Maclura pomifera

  20b. Shrubs; leaves and twigs without milky sap...................... *JAPANESE BARBERRY*
                   Berberis thunbergii

 19b. Spines or thorns absent.
  21a. Pith divided by woody plates; some leaves with
    short abrupt tip.. *BLACK GUM*
                   Nyssa sylvatica

  21b. Pith solid; leaves without short abrupt tips.
   22a. Each bud covered by a single hoodlike scale .... WILLOW
                   Salix spp.

   22b. Each bud covered by two or more scales; ends of secondary
     veins curving parallel with the leaf margins.............................. BUCKTHORN
                   Rhamnus spp.

4b. Margin of leaves with lobes or teeth.
 23a. Leaves distinctly lobed.
  24a. Leaves deeply palmately divided into about seven to nine lobes...*FATSIA*
                   Fatsia japonica

 24b. Leaves not as above.
  25a. Large, shallowly V-shaped notch at the apex of each leaf; fine line
    (stipule scar) encircling twig at each node .... *TULIP TREE*
                   Liriodendron tulipifera

  25b. Leaves and stipule scars not as above.
   26a. Leaves pinnately veined.
    27a. Leaves exuding milky sap when broken off twig.
     28a. Lobes with several teeth .... *MULBERRY*
                   Morus spp.

     28b. Lobes sometimes wavy-margined but without distinct teeth .... *FIG*
                   Ficus carica

    27b. Leaves without milky sap.
     29a. Lobes with pointed teeth.
      30a. Teeth confined to ends of lobes... *BLACK OAK GROUP*
                   Quercus spp.

      30b. Teeth along entire margin.. *CRABAPPLE, HAWTHORN*
                   Crataegus spp.

     29b. Lobes rounded without teeth .... *WHITE OAK GROUP*
                   Quercus spp.

   26b. Leaves palmately veined.
    31a. Leaves exuding milky sap from petiole when broken off twig.
     32a. Lobes with several teeth .... *MULBERRY*
                   Morus spp.

32b. Lobes sometimes wavy-margined but without distinct teeth .... *FIG*
Ficus carica

31b. Leaves without milky sap.
33a. Vines.
34a. Climbing by means of short, branched tendrils with disks .... *BOSTON IVY*
Parthenocissus tricuspidata

34b. Climbing by means of short adventitious roots at right
angles to the stems .... *ENGLISH IVY*
Hedera helix

33b. Erect trees or shrubs.
35a. Base of petiole hollow and enclosing the axillary bud .... *SYCAMORE*
Platanus spp.

35b. Base of petiole not enclosing axillary bud.
36a. Lower surface of leaves white-woolly .... *SILVER POPLAR*
Populus alba

36b. Lower surface of leaves not white-woolly.
37a. Older bark shredding easily from trunk or branches .... *NINEBARK*
Physocarpus opulifolius

37b. Older bark not readily separating from the trunk.
38a. Base of blade forming an acute angle with the petiole;
stipules present; shrubs .... *ROSE OF SHARON*
Hibiscus syriacus

38b. Base of blade not forming an acute angle with the
petiole; stipules absent; trees .... *SWEET GUM*
Liquidambar styraciflua

23b. Leaves toothed but not lobed.
39a. Thorns present on twigs and branches.
40a. Petioles with glands on upper surfaces .... *PLUM, CHERRY*
Prunus spp.

40b. Petioles without glands.
41a. Thorns leafy or with axillary buds, on branches only .... *WILD PEARS, APPLES*
Pyrus spp.

41b. Thorns without leaves or axillary buds, on branches and twigs.
42a. Leaves with large kidney-shaped stipules .... *JAPANESE QUINCE*
Chaenomeles lagenaria

42b. Leaves without stipules .... *CRABAPPLE, HAWTHORN*
Crataegus spp.

39b. Thorns absent.
43a. Climbing vines .... *BITTERSWEET*
Celastrus scandens

43b. Stems erect; trees or shrubs.
44a. Leaves with three or more main veins from near the base.
45a. Leaves exuding milky sap from petiole when broken off twig.................................. *MULBERRY*
Morus spp.

45b. Leaves without milky sap.
46a. Leaves noticeably asymmetrical at the base of the blade; petioles
more or less rounded in cross section.
47a. Leaves somewhat egg-shaped in outline and tapering toward the tip;
corky ridges present on branches .... *HACKBERRY*
Celtis spp.

47b. Leaves heart-shaped in outline; corky ridges absent .... *BASSWOOD, LINDEN*
Tilia spp.

46b. Leaves not noticeably asymmetrical at base of blade.
    48a. Petioles somewhat flattened; leaves roughly
        as wide as long .... *ASPEN, POPLAR, COTTONWOOD*
Populus spp.

    48b. Petioles not flattened; leaves longer than wide .... *CEANOTHUS*
Ceanothus spp.

44b. Leaves with a single main vein from the base.
  49a. Buds on distinct stalks.
    50a. Buds somewhat sticky; leaf margins with slightly irregular double row of
        indistinct teeth (i.e., the larger teeth have smaller teeth along their margins)............... *ALDER*
Alnus spp.

    50b. Buds woolly; margins of leaves wavy .... *WITCH HAZEL*
Hamamelis virginiana

  49b. Buds not stalked.
    51a. Each bud covered by a single hoodlike scale .... *WILLOW*
Salix spp.

    51b. Buds with two or more scales.
      52a. Tips of teeth curving inward and/or petioles flattened .... *POPLAR*
Populus spp.

      52b. Tips of teeth not curving inward; petioles not flattened.
        53a. Leaves four to six times as long as broad.
          54a. Leaves curling back lengthwise, especially toward the tip .... *PEACH, ALMOND*
Prunus spp.

          54b. Leaves not curling back; each secondary
             vein with a hair-pointed tooth at the tip .... *AMERICAN CHESTNUT*
Castanea dentata

        53b. Leaves less than four times as long as broad.
          55a. Petioles with one or more pairs of glands on upper surface .... *PLUM, CHERRY*
Prunus spp.

          55b. Petioles without glands.
            56a. Thorns present on branches and twigs .... *CRABAPPLE, HAWTHORN*
Crataegus spp.

            56b. Thorns absent.
              57a. Buds clustered at tips of twigs; fruit an acorn .... *OAK*
Quercus spp.

              57b. Buds not clustered at tips of twigs; fruit not an acorn.
                58a. Leaves with a double row of teeth (i.e., with small teeth
                  along the margins of larger teeth).
                  59a. Base of leaf blade asymmetrical; trees .... *ELM*
Ulmus spp.

                  59b. Base of leaf blade symmetrical; shrubs.
                    60a. Twigs green, ridged .... *KERRIA*
Kerria japonica

                    60b. Twigs brown, not ridged .... *HAZELNUT*
Corylus spp.

                58b. Leaves with a single row of teeth.
                  61a. Buds long and sharp-pointed .... *BEECH*
Fagus spp.

61b. Buds not as in 61a.
  62a. Secondary veins curving.
    63a. Young leaves usually woolly on the lower
      surface ....*APPLE*
      Malus X domestica

    63b. Leaves not woolly on the lower surface.
      64a. Leaves less than 5 cm (2 in.) long;
        shrubs ....*CEANOTHUS*
        Ceanothus spp.

      64b. Leaves more than 5 cm (2 in.) long; trees .... *PEAR*
        Pyrus communis

  62b. Secondary veins not curving.
    65a. Mature bark peeling in papery sheets; lenticels conspicuous
      as horizontal lines....................................... *BIRCH*
      Betula spp.

    65b. Bark and lenticels not as above.
      66a. Leaves leathery; teeth ending in a stiff spine ....*HOLLY*
      Ilex spp.

      66b. Leaves not leathery; teeth without spines and
        confined to the upper ¾ of the blade.................*SPIRAEA*
        Spiraea spp.

13b. Lower leaves on a twig usually opposite;
  petioles very short or absent ....*CRAPE MYRTLE*
  Lagerstroemia indica

2b. Leaves compound.
  67a. Vines.
    68a. Plant climbing by means of branched tendrils with discs.
      69a. Leaves with three leaflets .... *BOSTON IVY*
      Parthenocissus tricuspidata

      69b. Leaves palmately compound with five leaflets ....*VIRGINIA CREEPER*
      Parthenocissus quinquefolia

    68b. Plant climbing without branched tendrils; leaves pinnately compound
      with nine to nineteen leaflets ....*WISTERIA*
      Wisteria spp.

  67b. Erect trees or shrubs.
    70a. All leaves one-pinnately compound.
      71a. Epidermal prickles or thorns present.
        72a. Stipules fused to lower half of petiole .... *ROSE*
        Rosa spp.

        72b. Stipules absent ....*BLACKBERRY, RASPBERRY*
        Rubus spp.

      71b. Epidermal prickles or thorns absent.
        73a. Leaflets entire and rounded or slightly notched at the tip, blue-green in color;
          stipules usually modified as spines .... *BLACK LOCUST*
          Robinia pseudo-acacia

        73b. Leaves not as above in all respects.
          74a. Leaflets with stiff marginal spines .... *OREGON GRAPE*
          Berberis aquifolium

        74b. Leaflets without spines.
          75a. Leaflets rank-smelling when crushed; gland present on lower surface
            of the few teeth that may be present on some of the lower leaflets ....*TREE OF HEAVEN*
            Ailanthus altissima

75b. Leaflets not as in 75a.
    76a. Rachis bearing two leaflets at the tip..*PISTACHIO*
        Pistacia chinensis

    76b. Rachis bearing a single leaflet at the tip.
        77a. Terminal leaflets similar in size to smaller than leaflets below.
            78a. Leaflets nine to seventeen, distinctly toothed .... *MOUNTAIN ASH*
                Pyrus spp.

            78b. Leaflets 15–23, obscurely toothed .... *BLACK WALNUT*
                Juglans nigra

        77b. Terminal leaflets distinctly larger than leaflets below.
            79a. Leaflets without teeth.. *ENGLISH WALNUT*
                Juglans regia

            79b. Leaflets toothed.
                80a. Buds with two to three pairs of scales.
                    81a. Leaflets five to eleven; winter buds bright yellow .... *BITTERNUT HICKORY*
                        Carya cordiformis

                    81b. Leaflets 11–17; winter buds yellow-brown ....*PECAN*
                        Carya illinoensis

                80b. Buds with more than six overlapping scales.
                    82a. Bark shaggy; leaflets usually five.. *SHAGBARK HICKORY*
                        Carya ovata

                    82b. Bark not shaggy; leaflets seven to nine .... *MOCKERNUT HICKORY*
                        Carya tomentosa

70b. All or some of the leaves bipinnately compound.
    83a. Leaflets toothed or lobed to varying degrees; blades mostly only partially
        dissected into secondary leaflets... *GOLDENRAIN TREE*
        Koelreuteria paniculata

    83b. Leaflets not as above.
        84a. Leaflets 2 cm (0.8 in.) or less long.. *ACACIA, WATTLE*
            Acacia spp.

        84b. Leaflets more than 1 cm (0.4 in.) long.
            85a. Leaflets more than twice as long as broad; branching spines sometimes
                present on the trunk...*HONEY LOCUST*
                Gleditsia triacanthos

            85b. Leaflets less than twice as long as broad; spines never present .... *KENTUCKY COFFEE TREE*
                Gymnocladus dioica

1b. Leaves opposite or whorled.
  86a. Leaves opposite.
    87a. Leaves simple.
      88a. Leaves palmately veined and lobed.
        89a. Petioles with stipules and glands .... *VIBURNUM*
            Viburnum spp.

        89b. Petioles without stipules and glands.
            90a. Leaves deeply lobed.
                91a. Leaves whitish on lower surface; inner surface of peeled bark salmon-colored .... *SILVER MAPLE*
                    Acer saccharinum

                91b. Leaves not whitish on lower surface; inner surface of peeled bark not salmon-colored.
                    92a. Leaves mostly more than 10 cm (4 in.) wide; lobes essentially
                      without teeth ....*BIG LEAF MAPLE*
                    Acer macrophyllum

92b. Leaf blades mostly less than 10 cm (4 in.) wide; margins of
   lobes toothed...*JAPANESE MAPLE*
   Acer palmatum

90b. Leaves shallowly lobed.
   93a. Leaves with milky sap ....*NORWAY MAPLE*
   Acer platanoides

   93b. Leaves without milky sap.
      94a. Leaves mostly with five to seven lobes .... *SUGAR MAPLE*
      Acer saccharum

      94b. Leaves mostly with three lobes.. *RED MAPLE*
      Acer rubrum

88b. Leaves pinnately veined.
   95a. Leaves mostly 12–30 cm (4.75–12.0 in.) long.
      96a. Leaves entire, hairy; large trees.....................................*ROYAL PAULOWNIA*
      Paulownia tomentosa

      96b. Leaves toothed, without hairs; shrubs.................................*HYDRANGEA*
      Hydrangea macrophylla

   95b. Leaves mostly less than 12 cm (4.75 in.) long.
      97a. Stems climbing.. *HONEYSUCKLE*
      Lonicera spp.

      97b. Stems erect, not climbing.
         98a. Leaves spicy-aromatic when crushed................................ *WESTERN SPICEBUSH*
         Calycanthus occidentalis

         98b. Leaves not spicy-aromatic when crushed.
            99a. Leaf margins toothed.
               100a. Axillary buds visible.
                  101a. Bud scales distinctly in pairs or absent ....*VIBURNUM*
                  Viburnum spp.

                  101b. Bud scales overlapping, not in pairs.
                     102a. Stipules present; leaf margins with double row of fine teeth .... *JETBEAD*
                     Rhodotypos kerrioides

                     102b. Stipules absent; leaf margins with single row of teeth.
                        103a. Pith hollow or with transverse plates...*FORSYTHIA*
                        Forsythia spp.

                        103b. Pith solid, without transverse plates ....*WAHOO*
                        Euonymus spp.

               100b. Axillary buds hidden by base of petiole .... *MOCK ORANGE*
               Philadelphus spp.

            99b. Leaf margins essentially entire (without teeth).
               104a. Leaves with secondary veins curving parallel with margin or turning
                  back to fuse with other secondary veins.
                  105a. Leaves mostly less than 3 cm (1.25 in.) long; secondary veins forming
                     a network .... *ABELIA*
                     Abelia spp.

                  105b. Leaves more than 5 cm (2 in.) long; secondary veins curving but free
                     at their ends ....*DOGWOOD*
                     Cornus spp.

104b. Leaves not as in 104a.
    106a. Leaves heart-shaped and tapering to a point .... *LILAC*
        Syringa vulgaris

    106b. Leaves roughly egg-shaped in outline, not tapering to a point ....*PRIVET*
        Ligustrum spp.

87b. Leaves compound.
  107a. Stems climbing.
    108a. Leaflets two or three.
      109a. Leaflets two..*TRUMPET FLOWER*
        Bignonia capreolata

      109b. Leaflets three... *CLEMATIS*
        Clematis spp.

    108b. Leaflets seven to eleven .... *TRUMPET VINE*
        Campsis radicans

  107b. Stems erect; trees or shrubs.
    110a. Leaves palmately compound.
      111a. Leaflets mostly five.. *BUCKEYE*
        Aesculus spp.

      111b. Leaflets mostly seven .... *HORSE CHESTNUT*
        Aesculus hippocastanum

    110b. Leaves pinnately compound.
      112a. Leaflets mostly three to five, a few large teeth along the margins.........................*BOX ELDER*
        Acer negundo

      112b. Leaflets mostly five to seven, margins entire to regularly toothed.
        113a. Pith small; lenticels inconspicuous; trees...*ASH*
          Fraxinus spp.

        113b. Pith large; lenticels raised and conspicuous; shrubs ....*ELDERBERRY*
          Sambucus spp.

86b. Leaves in whorls of three..*CATALPA*
    Catalpa spp.

A Key to Some Common Deciduous Trees and Vines in Their Winter Condition

1a. Leaf scars alternate or in a spiral.
  2a. Stem erect.
    3a. Bundle scars three or more in a V-shaped or crescent-shaped line.
      4a. Stipules or stipule scars present.
        5a. Terminal bud present.
          6a. Pith circular in outline; bud scales sticky; buds not stalked...*COTTONWOOD*
            Populus spp.

          6b. Pith triangular in outline; bud scales with feltlike surface; buds on short stalks .... *ALDER*
            Alnus spp.

        5b. Terminal bud absent.
          7a. Axillary bud not completely surrounded by leaf scar.
            8a. One hoodlike scale covering each bud ....*WILLOW*
              Salix spp.

            8b. Several overlapping scales covering each bud.
              9a. Older bark of tree peeling in thin papery sheets; lenticels forming horizontal lines........... *BIRCH*
                Betula spp.

9b. Bark and lenticels not as in 9a.
　10a. Buds definitely lopsided in appearance, usually red, occasionally
　　greenish .... *BASSWOOD, LINDEN*
　　　　　Tilia spp.

　10b. Buds more symmetrical in appearance, brown in color.
　　11a. Tree trunk and older branches with more or less vertical corky ridges;
　　　tips of buds flattened; pith with transverse plates .... *HACKBERRY*
　　　　　Celtis occidentalis

　　11b. Bark without raised corky ridges; tips of buds not flattened;
　　　pith without transverse plates .... *ELM*
　　　　　Ulmus spp.

7b. Axillary bud completely surrounded by leaf scar .... *SYCAMORE*
　　　Platanus spp.

4b. Stipules or stipule scars absent.
　12a. Terminal bud present.
　　13a. Pith chambered.
　　　14a. Bark of tree brown, rough..*BLACK WALNUT*
　　　　　Juglans nigra

　　　14b. Bark of tree gray, smooth (except in old trees near the base)..............................*ENGLISH WALNUT*
　　　　　Juglans regia

　　13b. Pith not chambered.
　　　15a. Bundle scars three.
　　　　16a. Branches often with "pie crust" corky ridges of bark; bud scales with
　　　　　tiny hairs along the margins.. *SWEET GUM*
　　　　　　Liquidambar styraciflua

　　　　16b. Branches without corky ridges; bud scales without hairs along
　　　　　the margins... *PLUMS, CHERRIES*
　　　　　　Prunus spp.

　　　15b. Bundle scars more than three.
　　　　17a. Buds yellowish brown..*SUMAC*
　　　　　　Rhus spp.

　　　　17b. Buds dark reddish brown .... *PEAR*
　　　　　　Pyrus spp.

　12b. Terminal bud absent.
　　18a. Bundle scars three.
　　　19a. Twigs angular in cross section; stipular spines often present ....*BLACK LOCUST*
　　　　　Robinia pseudo-acacia

　　　19b. Twigs more or less round in cross section; stipular spines absent but large
　　　　branched spines sometimes present on the trunk ....*HONEY LOCUST*
　　　　　Gleditsia triacanthos

　　18b. Bundle scars five or more.
　　　20a. Leaf scars V-shaped, partly surrounding the axillary bud..*SUMAC*
　　　　　Rhus spp.

　　　20b. Leaf scars semicircular or triangular.
　　　　21a. Pith yellowish tan; twigs rank-smelling when bruised ....*TREE OF HEAVEN*
　　　　　　Ailanthus altissima

　　　　21b. Pith salmon-colored; twigs not rank-smelling when bruised........... *KENTUCKY COFFEE TREE*
　　　　　　Gymnocladus dioica

3b. Bundle scar one (or seemingly one), arranged in a circle or irregularly scattered.
22a. Bundle scars four to many, arranged in a circle or irregularly scattered.
    23a. Terminal bud present.
        24a. Stipule scars barely extending back from upper corners of leaf scars .... *OAK*
                Quercus spp.

        24b. Stipule scars partially or completely encircling twig.
            25a. Bud scales several, overlapping.
                26a. Bundle scars scattered; buds long-pointed; milky sap not present .... *BEECH*
                      Fagus spp.

                26b. Bundle scars in a circle; buds oval in shape; milky sap present in twigs .... *FIG*
                      Ficus spp.

            25b. Bud scales one or two, forming a hood.
                27a. Bark smooth, gray; bud scale one; leaf scars not round .... *JAPANESE MAGNOLIA*
                      Magnolia soulangeana

                27b. Bark somewhat roughened, brownish; bud scales two; leaf scars roundish .... *TULIP TREE*
                      Liriodendron tulipifera

    23b. Terminal bud absent.
        28a. Visible bud scales four or more.. *MULBERRY*
                Morus spp.

        28b. Visible bud scales two or three.
             29a. Buds and twigs light olive brown; twigs nearly straight .... *AMERICAN CHESTNUT*
                      Castanea dentata

             29b. Buds and twigs reddish, occasionally greenish; twigs zigzag in appearance ..... *BASSWOOD, LINDEN*
                      Tilia spp.

22b. Bundle scar one or seemingly one.
    30a. Terminal bud absent; twigs gray, not aromatic when bruised.. *PERSIMMON*
        Diospyros spp.

    30b. Terminal bud present; twigs green, aromatic when bruised .... *SASSAFRAS*
        Sassafras spp.

2b. Stem climbing or twining.
  31a. Stems with aerial roots or tendrils present; buds not silky-haired.
    32a. Outer bark shreddy and peeling easily; pith with woody partitions at the nodes .... *GRAPE*
        Vitis spp.

    32b. Outer bark not peeling easily; pith without woody partitions at the nodes.
        33a. Stems with aerial roots only..*POISON IVY, POISON OAK*
                Toxicodendron spp.

        33b. Stems with tendrils only..*BOSTON IVY, VIRGINIA CREEPER*
                Parthenocissus spp.

  31b. Stems without aerial roots or tendrils present; buds silky-haired .... *WISTERIA*
        Wisteria sinensis

1b. Leaf scars opposite or whorled.
  34a. Stem climbing...*TRUMPET FLOWER*
        Bignonia capreolata

34b. Stem erect.
  35a. Leaf scars whorled, mostly in threes ....*CATALPA*
        Catalpa spp.

35b. Leaf scars opposite.
    36a. Bundle scars tiny and numerous, forming U-shaped lines .... *ASH*
        Fraxinus spp.

    36b. Bundle scars distinct and separate.
        37a. Bundle scars more than 12, arranged in an ellipse .... *ROYAL PAULOWNIA*
            Paulownia tomentosa

        37b. Bundle scars less than 12, not arranged in an ellipse.
            38a. Bundle scars three.
                39a. Junction of leaf scars forming raised points at right angles to the whitish axillary buds ... *BOX ELDER*
                    Acer negundo

                39b. Leaf scars not as above; buds not whitish.
                    40a. Terminal buds large and conspicuous; axillary buds minute and barely visible
                        to the naked eye ....*DOGWOOD*
                        Cornus spp.

                    40b. Terminal buds slightly larger than the easily seen axillary buds .... *MAPLE*
                        Acer spp.

            38b. Bundle scars five or more.
                41a. Terminal bud absent; pith occupying at least $^1/_3$ the diameter of the twig .... *ELDERBERRY*
                  Sambucus spp.

                41b. Terminal bud present; pith small.. *BUCKEYE, HORSE-CHESTNUT*
                  Aesculus spp.

A Key to Some Major Western Conifers

1a. "Fruit" berrylike.
    2a. Leaves narrow and flattened, about 1–5 cm (0.4–2.0 in.) long.
        3a. "Fruit" red; tree usually near streams..*WESTERN YEW*
            Taxus brevifolia

        3b. "Fruit" greenish; leaves spine-tipped .... *CALIFORNIA NUTMEG*
            Torreya californica

    2b. Leaves very tiny, scalelike and overlapping, completely covering the twigs, aromatic;
        "fruit" blue or black...*JUNIPER*
        Juniperus spp.

1b. "Fruit" a woody cone.
    4a. Cones with scales that overlap like shingles.
        5a. Leaves scalelike, tiny; branches in flattened sprays .... *INCENSE CEDAR*
            Calocedrus decurrens

        5b. Leaves narrow and flattened, or needlelike.
            6a. Leaves narrow and flattened.
                7a. Cones borne erect on branches, disintegrating easily at maturity; thin woody bracts
                  not present between cone scales.
                  8a. Leaves with a twist on a short petiole .... *WHITE FIR*
                    Abies concolor

                  8b. Leaves sessile (without petioles)..*RED FIR*
                    Abies magnifica

                7b. Cones pendent, falling whole from the tree; thin woody bracts present between
                the cone scales.
                  9a. Cones usually less than 2 cm (0.75 in.) long; leaves appearing two-ranked ....*WESTERN HEMLOCK*
                    Tsuga mertensiana

9b. Cones usually more than 2 cm (0.75 in.) long; leaves arranged spirally around the twig.
    10a. Leaves definitely flattened; buds pointed; cone bracts longer than the scales .... *DOUGLAS FIR*
          Pseudotsuga menziesii

    10b. Leaves stiff and four-angled or somewhat flattened; buds rounded; cone bracts
        shorter than the scales... *SPRUCE*
          Picea spp.

6b. Leaves needlelike.
    10a. Two needles, each less than 7.5 cm (3 in.) long, in a cluster ....*LODGEPOLE PINE*
          Pinus murrayana

    10b. Three needles (occasionally two), each 7.5–20.0 cm (3–8 in.) long, or five
      (occasionally four) needles in a cluster, each less than 10 cm (4 in.) long.
      11a. Three needles in a cluster.
        12a. Cones remaining closed and attached to tree for years..*KNOBCONE PINE*
            Pinus attenuata

        12b. Cones opening and falling when mature.
          13a. Foliage gray-green and thin; cones heavy, the cone scales curving back
            at their tips .... *GRAY PINE*
              Pinus sabiniana

          13b. Foliage green and relatively thick; cones and scales not as above.
            14a. Cones 7.5–12.5 cm (3–5 in.) long..*YELLOW PINE*
               Pinus ponderosa

            14b. Cones 12.5–25.0 cm (6–10 in.) long...*JEFFREY PINE*
               Pinus jeffreyi

      11b. Five needles in a cluster.
        15a. Cones less than 30 cm (12 in.) long.
          16a. Cones 2.5–7.5 cm (1–3 in.) long; cones globose; high-altitude trees.............. *WHITEBARK PINE*
            Pinus albicaulis

          16b. Cones 15–20 cm (6–8 in.) long; cones slender; intermediate-altitude trees............. *SILVER PINE*
            Pinus monticola

         15b. Cones more than 30 cm (12 in.) long .... *SUGAR PINE*
          Pinus lambertiana

4b. Cones whose scales, with the exception of the broad flattened tips, do not overlap like shingles.
  17a. Center of scale tips depressed.
    18a. Leaves narrow and flattened; cones 1.0–2.5 cm (0.4–1.0 in.) long .... *COASTAL REDWOOD*
          Sequoia sempervirens

    18b. Leaves small, short, and pointed; cones 5–10 cm (2–4 in.) long................................*GIANT SEQUOIA*
          Sequoiadendron giganteum

  17b. Center of scale tips not depressed.
    19a. Center of scale tip ending in a sharp point; cones about 1–2 cm (0.40–0.75 in.)
      in diameter; foliage gray-green ....*MACNAB CYPRESS*
          Cupressus macnabiana

    19b. Center of scale tip not sharply pointed; cones about
      0.75 cm (0.25 in.) in diameter.. *PORT ORFORD CEDAR*
          Chamaecyparis lawsoniana

A Key to Some Major Eastern and Southern Conifers

1a. "Fruit" berrylike.
  2a. Leaves narrow and flattened, about 1–5 cm (0.4–2.0 in.) long; "fruit" red .... *YEW*
          Taxus canadensis

2b. Leaves very tiny, scalelike, and overlapping, completely covering the twigs.
   3a. Erect trees ....RED CEDAR
        Juniperus virginiana

   3b. Prostrate shrubs ....*JUNIPER*
        Juniperus spp.

1b. "Fruit" a woody cone.
   4a. Cones with scales that overlap like shingles.
      5a. Leaves tiny, scalelike, and overlapping; branches in flattened sprays............................. *ARBOR-VITAE*
         Thuja spp.

      5b. Leaves narrow and more or less flattened, or needlelike.
      6a. Leaves narrow and more or less flattened.
        7a. Cones borne erect on branches, disintegrating at maturity ....*BALSAM FIR*
           Abies balsamea

        7b. Cones pendent, falling whole from the tree.
          8a. Cones usually less than 2 cm (0.75 in.) long; leaves definitely flattened
             and appearing two-ranked.. *EASTERN HEMLOCK*
             Tsuga canadensis

          8b. Cones usually more than 2 cm (0.75 in.) long; leaves stiff and four-angled
             or somewhat flattened, arranged spirally around the twig.. *SPRUCE*
             Picea spp.

      6b. Leaves needlelike.
        9a. Leaves in tight spirals toward tips of short spurlike branches; deciduous... *LARCH*
           Larix spp.

        9b. Leaves in clusters of two to five.
          10a. Leaves two to three per cluster.
            11a. Leaves mostly two per cluster.
               12a. Leaves 8–14 cm (3.25–5.50 in.) long .... *SHORTLEAF PINE*
                 Pinus echinata

              12b. Leaves 2–8 cm (0.75–3.25 in.) long.
                13a. Leaves 2–4 cm (0.75–1.50 in.) long; cones often remaining closed for years............. *JACK PINE*
                  Pinus banksiana

                13b. Leaves 4–8 cm (1.50–3.25 in.) long; cones normally opening.
                  14a. Cone scales with a prickle at the tip; leaves twisted .... *VIRGINIA PINE*
                    Pinus virginiana

                  14b. Cone scales without a prickle; leaves relatively straight...*RED PINE*
                    Pinus resinosa

           11b. Leaves mostly three per cluster.
               15a. Leaves 7–12 cm (3.00–4.75 in.) long .... *PITCH PINE*
                Pinus rigida

               15b. Leaves 15–45 cm (6–18 in.) long.
                16a. Cones 15–25 cm (6–10 in.) long, the scale prickles curving in........................*LONGLEAF PINE*
                  Pinus palustris

                16b. Cones mostly less than 15 cm (6 in.) long, the scale prickles not curving in.
                  17a. Cones shiny, on stalks .... *SLASH PINE*
                    Pinus caribaea

                  17b. Cones dull, without stalks...*LOBLOLLY PINE*
                    Pinus taeda

           10b. Leaves in clusters of five.. *EASTERN WHITE PINE*
            Pinus strobus

4b. Cones with wedge-shaped scales that do not overlap toward their bases...*BALD CYPRESS*
   Taxodium distichum

Laboratory Examination Sheet

NAME _____

LAB SECTION NO. _____

DATE _____

1. _____
2. _____
3. _____
4. _____
5. _____
6. _____
7. _____
8. _____
9. _____
10. _____
11. _____
12. _____
13. _____
14. _____
15. _____
16. _____
17. _____
18. _____
19. _____
20. _____
21. _____
22. _____
23. _____
24. _____
25. _____

26. _____
27. _____
28. _____
29. _____
30. _____
31. _____
32. _____
33. _____
34. _____
35. _____
36. _____
37. _____
38. _____
39. _____
40. _____
41. _____
42. _____
43. _____
44. _____
45. _____
46. _____
47. _____
48. _____
49. _____
50. _____

NAME _____

51. _____   76. _____

52. _____   77. _____

53. _____   78. _____

54. _____   79. _____

55. _____   80. _____

56. _____   81. _____

57. _____   82. _____

58. _____   83. _____

59. _____   84. _____

60. _____   85. _____

61. _____   86. _____

62. _____   87. _____

63. _____   88. _____

64. _____   89. _____

65. _____   90. _____

66. _____   91. _____

67. _____   92. _____

68. _____   93. _____

69. _____   94. _____

70. _____   95. _____

71. _____   96. _____

72. _____   97. _____

73. _____   98. _____

74. _____   99. _____

75. _____   100. _____

Appendix

Following are lists of the materials needed in the laboratory for use with *Stern's Introductory Plant Biology Laboratory Manual.* The materials listed will accommodate a laboratory section of 20 students.

These materials should be available throughout the term of the course: compound microscopes, dissecting microscopes, lens paper, tissues, blank microscope slides and coverslips, forceps, scissors, probes, single-edge razor blades, toothpicks, stick-on labels, straight pins, empty dropper bottles and eyedroppers, blank note cards, masking tape, stapler and staples, ball of string, grease pencil, box of Whatman filter paper circles (9 cm in diameter), plastic rulers, replacement bulbs for microscope lights, and adhesive bandages. Specific prepared microscope slides needed are the following:

1. *Allium* root rip, mitosis, longisection
2. *Tilia,* 2-year stem, cross section
3. *Medicago,* young and mature stem, cross section
4. *Zea mays,* stem, cross section
5. *Ranunculus,* mature root, cross section
6. *Smilax,* mature root, cross section
7. *Salix,* branch root origin, cross section
8. *Syringa,* leaf cross section
9. *Pinus,* leaf cross section
10. *Anabaena,* whole mount
11. *Oedogonium,* whole mount
12. *Spirogyra,* whole mount
13. *Ulothrix,* whole mount
14. *Rhizopus,* whole mount
15. *Penicillium,* whole mount
16. *Peziza,* apothecia, cross section
17. *Physcia,* thallus
18. *Coprinus,* pileus
19. Moss protonema
20. *Mnium,* antheridia, longisection
21. *Mnium,* archegonia, longisection
22. Fern prothallus
23. *Pinus,* archegonium, longisection
24. *Lilium,* anthers, cross section
25. *Lilium,* mature embryo sac, cross section

Exercise 1

The Microscope (students work individually)

1. Microscope slides on each of which is mounted a printed letter *e; 1 slide per student*
2. Crossed silk fiber slides; *1 slide per student*
3. Pond water;* *2 or 3 bowls or jars*

*The pond water should be as rich and varied as possible. It can be made so by including scrapings from sides of aquaria, water from sewage treatment ponds, material from seepage banks, sluggish water in sloughs, etc. Generally, try to avoid too much material of a dark bluish green color, as this often contains only one or two forms of cyanobacteria, which are much less interesting than eukaryotic forms to a beginning student.

Exercise 2

The Cell (students work individually)

1. Healthy *Elodea* sprigs; *4 bowls*
2. Fresh onion; *1*
3. White potato; *half a small potato*
4. Razor blades near potato and pepper; *3 or 4*
5. IKI solution; *2 dropper bottles*
6. *Tradescantia* plants with flowers;* *2 flats*
7. Beaker (400 ml) of tap water (labeled); *1*
8. Red pepper or tomato; *1 fresh slice*
9. Pond water (similar to that for Exercise 1); *1 bowl*
10. Eyedroppers for tap water; *2*
11. Plant cell models and charts

*Tradescantia paludosa and other *Tradescantia* spp. usually produce flowers that open in the morning and wither late the same day. If handled correctly, there are usually enough stamen hairs on one or two flowers to meet the needs of one laboratory section. For laboratories that meet late in the day or in the evening, stamen hairs from large, unopened buds will generally be satisfactory, particularly if the instructor has the student come to him or her with a drop of water on a slide and then strips hairs off individually with a fine-point forceps, placing the hair on the student's slide immediately after removing it from the flower. Epidermal hairs of *Gynura* (velvet plant) may be substituted if necessary.

Exercise 3

Mitosis (students work individually)

1. Squares of brown paper (approximately 60 cm × 60 cm); *enough for 20 students*
2. Modeling clay—2 colors;* *enough for 20 students*
3. Pipe cleaners—2 colors; *enough for 20 students*
4. Slides of mitosis in onion (*Allium*) root tip
5. Fresh onions suspended with toothpicks in water with growing roots; *2 per student*
6. Acetocarmine; *1 dropper bottle*
7. Models and/or charts of cells undergoing mitosis; *1 set*

**Remind the students not to mix the colors of modeling clay when they are finished with the exercise.

Exercise 4

Roots (students work individually)

1. Radish seedlings in petri dishes;* *20 seedlings*
2. Prepared slides of cross sections of young buttercup (*Ranunculus*) and greenbrier (*Smilax*) roots
3. Prepared slides of willow (*Salix*) roots showing lateral roots

**These are for examination of root hairs by the students. Remind students to discard material after removing tips.

Exercise 5

Stems (students work individually)

1. Live woody grape twigs, approximately 1 cm thick; *4 segments about 1 dm long*
2. Live basswood twigs, approximately 1 cm thick; *4 segments about 1 dm long and 2 to 3 years old*
3. Live *Begonia* or *Coleus* stems, approximately 1 cm thick; *4 segments about 1.5 dm long*
4. New razor blades; *5 or 6*
5. Live 2-year-old basswood or linden (*Tilia*) twigs
6. Watch glasses; *4*
7. Pure water; *1 dropper bottle*
8. Paraffin; *1 cake*

Each student laboratory table should have 1 set of dropper bottles of

1. Gentian violet*
2. Eosin solution**
3. Balsam or *Permount* in xylene
4. Phoroglucinol stain***
5. Clove oil
6. Xylene
7. 95% ethyl alcohol
8. Whatman filter paper circles (9 cm in diameter); *1 box*
9. Stock bottle of 95% ethyl alcohol; *1 liter*
10. Charts and models of both dicot and monocot stems; *1 set*

*Formula for gentian violet: 1 gm gentian violet, 20 ml 95% EtOH. Mix. Add 80 ml 5% formalin (95 ml distilled water + 5 ml commercial formalin). Mix well and dispense.

**Formula for eosin solution: 0.5 gm eosin in 100 ml 95% EtOH. Mix and dispense.

***Formula for phoroglucinol stain: 1 gm phoroglucinol in 100 ml 95% EtOH. Mix and dispense.

Exercise 6

Leaves (students work individually)

1. Twigs with simple leaves, pinnately compound leaves, palmately compound leaves, and leaves with stipules
2. Fresh monocot leaves
3. Display of insectivorous and other modified leaves
4. Healthy *Sedum* or *Zebrina* plant; *1*
5. Toothpicks; *1 box*
6. Prepared slides of cross sections of lilac (*Syringa*) and pine (*Pinus*) leaves
7. Prepared microscope slides of hydrophyte versus xerophyte leaves
8. Charts and models of roots and leaves; *1 set*

Exercise 7

Plant Propagation (students work in pairs or teams, except with African violet leaf cuttings and Begonia leaf cuttings, which are done individually)

1. Willow or poplar twigs, approximately 1.5 cm in diameter;* *20*
2. Willow or poplar twigs, approximately 0.75 cm in diameter;* *20*
3. Plastic or wooden flats with potting soil; *4*
4. *Rootone* or similar IAA powder; *adequate*
5. Plastic bags ("baggie" size) containing vermiculite; *20*
6. 1.5-dm long segments of *Begonia* or any other easily rooted stems; *20*
7. Rose cuttings; *4*
8. Carrots; *6*
9. White potatoes; *2*
10. Sweet potatoes; *2*
11. Grafting wax/sealing compound; *adequate*
12. Tissues; *2 boxes*
13. Wide-mouthed mason jars (or equivalent); *2*
14. Knife; *1*
15. Petri dishes; *6*
16. Toothpicks; *1 box*
17. African violets or *Peperomia* plants;** *adequate*
18. Enamel pans (for African violets);** *adequate*
19. Aluminum foil; *1 roll*
20. Bulbs (e.g., *Lilium*); *2*
21. Lily plant with axillary bulbils; *1*
22. Pineapple with top; *1*
23. Clay pots with potting soil; *2*
24. 10% liquid bleach solution; *500 ml*

25. Ripe apples; *2*
26. Long forceps; *1*
27. Nutrient agar slants; *2*
28. Test-tube rack; *1*
29. Bunsen burner; *1*
30. Matches; *1 box*
31. Flat rubber strips or plastic tape (for grafts); *adequate*
32. Single-edged razor blades; *adequate*
33. Melon-ball scoop; *1*
34. Light banks;*** *adequate*
35. Vermiculite; *1-peck sack*
36. Containers for bulbs; *1 or 2*
37. Several healthy *Equisetum* stems
38. Raw peanuts
39. Charts and models

*Bases of willow or poplar twigs should be kept in water both before and after grafting.
**African violet plants used in this exercise should have more than 20 leaves so that a few remain on the plant after each student has removed one. Insert petioles of leaves through small holes punched in aluminum foil spread over pans of water as shown. Be certain to space the holes so that the leaf blades do not overlap.

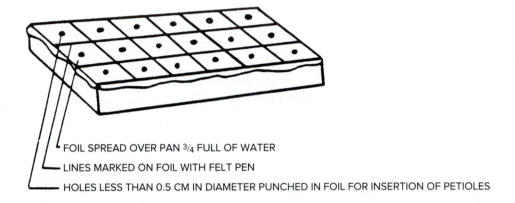

FOIL SPREAD OVER PAN ³/₄ FULL OF WATER
LINES MARKED ON FOIL WITH FELT PEN
HOLES LESS THAN 0.5 CM IN DIAMETER PUNCHED IN FOIL FOR INSERTION OF PETIOLES

***Unless a greenhouse is available for this exercise, it is strongly recommended that the cuttings, etc., be placed under light banks with a 24-hour timer in the line so that the lights are on for 16 hours and off for 8 hours. Students should be reminded to provide adequate light for the cuttings they take with them from the laboratory and to see that their projects receive sufficient water.

Exercise 8

Cell Components and Products (students work in teams)

1. Starch solution;* *250 ml*
2. Glucose solution;* *250 ml*
3. Protein solution;* *250 ml*
4. Thin Brazil nut slices floating in water;* *20*
5. Solutions of unknowns;* *1 set in 250-ml bottles*
6. Forceps; *2*
7. 400-ml beakers; *1 per table*
8. 50-ml beakers; *2 per table*
9. Petri dishes; *1 per table*
10. Test tubes; *9 per table*
11. Test-tube holders; *1 per table*
12. Mortar and pestle; *1 per table*
13. Hot-water baths; *1 per table*
14. Tripods and insulated wire gauze for hot-water baths; *1 set per table*
15. Test-tube brushes and dishpan with soapy or detergent water; *1 set*
16. IKI reagent for starch; *1 dropper bottle*
17. Protein alkaline solution; *250 ml*

18. Tes-tape (for glucose test); *1 roll*
19. Biuret reagent (2); *1 dropper bottle each*
20. Sudan IV (lipophilic stain); *1 dropper bottle*
21. Dilute HCl or vinegar solution; *1 dropper bottle*
22. Dilute NaOH or ammonium hydroxide solution; *1 dropper bottle*
23. Bunsen burner; *1 per table*
24. Matches
25. Pots with lids and racks; *1 1-gallon or larger pot per table*
26. Trays of ice cubes; *1 tray per table*
27. Facial tissues; *1 box*
28. Filter paper
29. 95% ethyl alcohol
30. Mint or other leaves with aromatic oils; *1 loose peck sack per table*
31. Head of purple or red cabbage; *1 per class*

*Prepare as follows:

Starch solution: Stir 1 gm soluble starch in a small amount of cold water until it is evenly suspended; then, while stirring, pour the mixture slowly into 100 ml of actively boiling water. Allow to boil for a few minutes and then dilute with 1 liter of distilled water. Test with Benedict's reagent to be sure it is free of reducing sugar.

Glucose solution: Dissolve 5 gm glucose in 1 liter of distilled water.

Protein solution: Add 10 gm powdered albumen per liter of distilled water and stir overnight with magnetic stirrer or until dissolved. Dilute 1:1.

Brazil nut slices: Cut with a rotary microtome set at 50 micrometers. The tissue of some nuts crumbles more easily than that of others; test before repeated slicing. The slices should be stained with a drop of Sudan IV; a single slice may be set up as a demonstration, if desired.

Unknown solutions: Combine the following from the stock solutions and dispense in numbered 250-ml bottles only:

glucose
glucose and starch
glucose and protein
starch and protein
glucose, starch, and protein
starch
protein
distilled water

Exercise 9

Diffusion, Growth, and Hormones (students work in teams)

1. India ink (dilute 1:1 aqueous); *2 dropper bottles*
2. Agar in petri dishes (for crystal dye diffusion); *2*
3. Potassium permanganate crystals; *several*
4. Green or blue dye crystals; *several*
5. Slides and coverslips; *adequate*
6. 25% sucrose solution; *2 dropper bottles*
7. IKI solution; *2 dropper bottles*
8. Healthy *Elodea* in finger bowl; *3 or 4 sprigs*
9. Starch-free string; *small ball*
10. Scissors; *1 pair*
11. Matches; *1 box*
12. Tie-on plant tags; *adequate*
13. Grid markers and ink pads; *1 set*
14. Aqueous eosin or food coloring; *500 ml*
15. 600-ml beakers; *2*
16. IAA in lanolin;* *1 small jar*
17. Lanolin; *1 small jar*
18. Rootone; *adequate*

19. Toothpicks; *adequate*
20. Single-edged razor blades; *2*
21. *Coleus* plants; *6*
22. Dropper bottles with pure water, 0.1 microgram gibberellic acid solution, and 1.0 microgram gibberellic acid solution
23. *Osmosis demonstration:* 3 osmometers—one with pure water in water, one with 25% sucrose solution in water, and one with 25% sucrose solution in 25% sucrose solution. With most simple osmometers, it is best to lower them into beaker fluid at an angle to avoid trapping air and also to raise them above the fluids in the beakers at the end of each laboratory session.
24. *Diffusion demonstration:* A glass cylinder held by a clamp on a ringstand is suspended over an Erlenmeyer flask containing ammonium hydroxide. A strip of litmus paper extending the length of the glass cylinder is held in place by a stopper at the top end. The ammonia solution is kept stoppered until used. A meter or half-meter ruler should be available for measuring the diffusion as it progresses.
25. *Dialysis demonstration:* Soak a 1-dm length of flattened dialysis tubing in distilled water until it softens and can be formed into a tube. Tie off one end with starch-free string. Fill the bag with dialyzing solution (soluble starch 1 gm per liter + 20 gm glucose per liter of starch solution). Tie off the other end of the bag and wipe dry. Place in a 100-ml beaker of distilled water. Provide a dropper bottle of IKI reagent for checking presence of starch in the beaker. When multiple sections are involved, prepare a fresh demonstration for each section.
26. *Transpiration demonstration:* Have a potted plant whose soil surface is covered with foil under a bell jar and a similar plant not so enclosed for a control.
27. *Phototropism demonstration:* Place a potted plant under a box that has a slit cut out of one side so that the plant can receive illumination from that direction only. Leave the plant under the box for at least 2 weeks in advance of the demonstration.
28. *Gravitropism demonstration:* Lay an actively growing potted plant on its side a day or two in advance of the demonstration. (Obvious gravitropic responses occur in *Coleus* plants within a few hours after they have been laid on their side.)
29. *Etiolation demonstration:* Germinate two sets of corn, bean, and pea seeds in pots about 2 weeks in advance of the laboratory session. Keep one set illuminated and the other in the dark.
30. *Ethylene gas demonstration:* Set up two bell jars under which have been placed two sprigs of holly with their bases in beakers of water. Include a ripe apple under one of the jars. Within 1 week, the ethylene from the ripening apple will have triggered abscission of the holly leaves.
31. *Turgor movements demonstration:* Have at least one healthy plant of *Mimosa pudica* and any available insectivorous plants with active traps (e.g., *Dionaea muscipula, Drosera* spp.) on display.
32. Balsam (*Impatiens*) plant; *1*
33. Potting soil and pots for planting; *adequate*

*Mix 2 mg IAA in 5 gm lanolin. Stir thoroughly with a glass rod. Keep in a capped jar. Prepare as close to time of use as possible, as its efficacy deteriorates rapidly.

Exercise 10
Photosynthesis (students work in teams)

1. Phenol red solution; *1 dropper bottle per table*
2. Forceps; *2*
3. Fresh *Elodea* sprigs for oxygen production; *4*
4. Fresh *Elodea* sprigs for carbon dioxide absorption; *2–3*
5. *Coleus* plants with leaves that have been partially covered with black paper for several days; *sufficient for 1 leaf for each student*
6. *Coleus* plants with leaves that have had normal exposure to light for several days; *sufficient for 1 leaf for each student*
7. 95% ethyl alcohol; *1 liter*
8. Petri dishes; *1 per table*
9. IKI solution; *1 dropper bottle*
10. 400-ml glass beakers; *1 per table*
11. Concentrated alcoholic solution of chlorophyll for fluorescence demonstration; *400 ml*
12. Freshly prepared green plant pigment extract;* *1 dropper bottle covered with foil and with a camel's-hair brush inserted in the stopper*
13. 1 × 15 cm paper chromatography strips; *25*
14. Rack containing stoppered, numbered test tubes, each with 1.5 cm of chromatography solvent;* *1 test tube per student*

15. Copper hot-water baths; *1 per table*
16. Tripods and insulated wire gauze for hot-water baths; *1 set per table*
17. Bunsen burners; *1 per table*
18. Test-tube brushes and dishpan with soapy or detergent water; *1 set*
19. Test tubes for phenol red solution; *1 set per table*
20. Test-tube holders; *1 per table*
21. Soda straws or glass tubing; *2 per table*

*Prepare as follows:

Plant pigment extract: Loosely fill a blender with ivy leaves. Add 100 ml acetone, and blend 20 seconds. Filter through cheesecloth and glass wool into a 250-ml separatory funnel. Allow the following to run slowly down the side of the funnel: 15 ml petroleum ether (will form a layer over the acetone) and 20 ml distilled water (to dilute acetone and force pigments into the petroleum ether). Allow to stand several minutes, and then drain off the acetone-water solution from below and discard. Collect the petroleum ether layer with the dissolved pigments for use in the exercise. Place in small vials and refrigerate until needed. Supply a fresh vial daily when multiple laboratory sections are involved. Prepare the pigment as close to the time of use as possible; it will not keep long.

Chromatography solvent: Mix 10 ml acetone with 90 ml petroleum ether.

Note: *Coleus* plants used in this exercise should have been well-illuminated for 2 or 3 days prior to use.

Exercise 11
Water in Plants; Respiration; Digestion (students work in teams)

1. Freshly-cut carnations; *3 per table*
2. Bottles of assorted food coloring; *3 per table*
3. Potometer; *1*
4. Small, fresh woody twig with broad leaves; *1*
5. Small electric fan; *1*
6. Barley seedlings under a glass jar; *1 pot containing about 20 seedlings*
7. Fresh potato or bean sprouts; *adequate*
8. Viable corn kernels; *30*
9. 10% liquid bleach; *1 pint*
10. Thermometers in stoppers or cotton plugs; *2*
11. Insulated bottles; *2*
12. Wide-mouthed glass bottles with stoppers; *2*
13. Glass tubing; *adequate*
14. Rubber tubing; *adequate*
15. 5% formalin solution; *1 liter*
16. Suspension of yeast in 10% sucrose solution; *1 dropper bottle*
17. Ripe apple; *1*
18. Baryta water (saturated solution of barium hydroxide in water); *1 liter*
19. Hydrogen peroxide; *1 dropper bottle*
20. Starch powder; *adequate*
21. Tetrazolium reagent;* *1 dropper bottle*
22. Syracuse watch glasses; *4*
23. Diastase; *adequate*
24. 400-ml beakers; *1 per table*
25. Bunsen burners; *1 per table*
26. Tripods; *1 per table*
27. Metal gauze (for tripods); *1 square per tripod*
28. Matches; *1 box*

*Tetrazolium reagent: Dissolve 1 gm 2, 3, 5-triphenyltetrazolium chloride in 100 ml of deionized water. Store in dark brown dropper bottles or cover with aluminum foil. Solution may turn pale yellow but remains active for about a week.

Exercise 12

Meiosis and Alternation of Generations (students work individually)

1. Modeling clay in 2 colors for centromeres; *adequate*
2. Pipe cleaners—2 lengths and 2 colors; *sufficient for 20 students*
3. Squares of brown paper (approximately 60 cm × 60 cm); *adequate*
4. Models of phases of meiosis; *1 set*

Exercise 13

Domain and Kingdom Survey (students work individually)

1. Agar plate with bacterial colonies; *2*
2. Cyanobacteria; *1 finger bowl or agar slant*
3. Green algae; *4 or 5 cultures*
4. *Euglena* culture; *1*
5. Dinoflagellates; *1 culture or portion*
6. Slime molds; *3 or 4*
7. True fungi, including morel, truffles, mushrooms, stinkhorn, bracket fungus, smut, rust; *1 set*
8. Herbarium sheets (or other preserved): club mosses, brown algae, red algae, *Ulva, Codium; 1 set*
9. Living specimens of mosses (with conspicuous sporophytes), liverworts, *Psilotum, Equisetum, Selaginella,* ferns, cycad, *Ginkgo,* conifers, flowering plants; *1 set*

Notes Concerning the Materials

Bacteria: Expose agar plates to house dirt, a doorknob, etc., 3 days in advance. Keep in the refrigerator after colonies develop; tape dishes shut and autoclave immediately after the exercise because of pathogens that could be present.

Cyanobacteria: These may be in the form of colonies collected at a pond (e.g., *Nostoc* balls), on agar slants, or growing on soil (e.g., *Oscillatoria*). Supplementary colored pictures of the live material would be helpful.

Green algae: Live colonies of *Hydrodictyon, Spirogyra, Scenedesmus, Volvox, Pediastrum,* desmids, and/or other similar material should be available.

Euglenoids: Sedimentation ponds of sewage treatment plants often are good sources of *Euglena* and *Phacus* if other sources are not available.

Dinoflagellates: These can be purchased from a biological supply house, or, if the institution happens to have access to an ocean during a red tide, excellent material can be obtained by scooping up small quantities of sea water and adding several drops of formalin. The organisms will be preserved indefinitely if treated in such fashion.

Slime molds: Both living and nonliving material is needed. *Physarum* plasmodia are easily cultured on agar in petri dishes; fruiting bodies of *Arcyria, Trichia, Stemonitis, Fuligo,* etc., are common on dead leaves, under damp logs, and in other places.

Exercise 14

Domains (Kingdoms) Archaea and Bacteria; Kingdom Protista (students work individually)

1. Petri plates of gram-positive and gram-negative bacteria; *2 each*
2. Bacterial plates showing a variety of colonies; *1 set*
3. Bunsen burners; *6*
4. Live and/or preserved *Anabaena* or *Nostoc* colonies
5. Living cultures of *Spirogyra, Oedogonium, Ulothrix, Volvox, Scenedesmus, Euglena,* or *Phacus,* diatoms; *1 set*
6. Slides of stained and preserved *Ulothrix, Spirogyra,* and *Oedogonium*
7. Diatomaceous earth; *small jar*
8. Herbarium specimens of *Gelidium, Porphyra, Gigartina, Ulva, Codium, Postelsia, Laminaria, Costaria, Nereocystis,* or *Desmarestia; 1 set*
9. Dinoflagellates; *adequate*
10. Loaf of sliced white bread containing no preservatives;* *adequate*
11. Petri dishes, small (plastic or glass); *1 per student*
12. Pond water;** *1 bowl*

13. Dropper bottles of gentian (crystal) violet dye (0.05% aqueous solution), safranin O dye,*** 95% EtOH and Gram's iodine reagent;**** *5 sets*
14. Living slime mold plasmodia (demonstration); *1 or 2 plates*
15. Nonexpendable and expendable slime mold sporangia;***** *adequate*
16. Specimens of slime molds (nonexpendable reproductive bodies); *several*

*Distribute bread to students (about 1/4 slice each); they should *barely* dampen it (more than a drop or two of water may promote growth of yeasts rather than filamentous fungi) and expose it for the development of fungal cultures for the next exercise.
**See note concerning pond water in Exercise 1.
***Formula for safranin O: 1 gm safranin O in 100 ml 95% EtOH.
****Formula for Gram's iodine reagent: 0.5 gm iodine and 1 gm KI in 150 ml distilled water.
*****There should be sufficient material for each student to dissect a few sporangia.

Exercise 15

Kingdom Fungi (Mycota) (students work individually)

1. Preserved, dried, or living specimens of puffballs, stinkhorns, earth stars, rusts, smuts, bracket fungi, bird's-nest fungi, other available fungi; *1 set*
2. Prepared slides of *Rhizopus, Penicillium, Peziza, Coprinus,* and *Physcia*
3. Lichens (crustose, foliose, fruticose); *1 set*

Exercise 16

Kingdom Plantae: Bryophytes and Ferns (students work individually)

1. Live mosses with sporophytes attached; *20*
2. Prepared slide of moss protonema
3. Live *Marchantia* with archegoniophores and antheridiophores
4. Petri dish with live protonemata (demonstration); *1*
5. Prepared slides of longitudinal sections of archegonial heads of *Mnium* (or similar moss)
6. Live hornworts (demonstration); *2 or 3*
7. Variety of ferns; *1 set*
8. Live prothalli (demonstration); *1 set*
9. Fern with expendable fronds that have mature sori; *adequate*

Exercise 17

Kingdom Plantae: Gymnosperms (students work individually)

1. Fresh pine branch with cluster of pollen cones (demonstration); *1*
2. Fresh conifer branches
3. Pine seed cones with seeds on scales; *2 or 3*
4. Conifer pollen (for examination under microscope); *small vial*
5. Prepared slides of longitudinal sections through a pine ovule
6. Cycad, *Ginkgo, Ephedra, Welwitschia, Gnetum; 1 set of live plants* (if available; alternatively a set of good pictures may be substituted)

Exercise 18

Kingdom Plantae: Angiosperms (Flowering Plants–Phylum Magnoliophyta) (students work individually)

1. Large flowers with superior ovaries;* *20*
2. Prepared slides of *Lilium*—cross sections of ovary showing embryo sacs; *Lilium*—cross sections of mature anthers
3. Flowers on display: a composite, a grass, a primitive flower (e.g., buttercup), an advanced flower with an inferior ovary (e.g., orchid), and an inflorescence; *1 set*
4. Flower model

*It is important that students have flowers that are large enough to be dissected. This enables them to discern the ovules in the ovary without too much difficulty and also to distinguish the calyx, corolla, and pistil easily.

Exercise 19

Fruits, Spices, and Survival Plants (students work individually)

1. As complete as possible representation of both fleshy and dry fruit types should be available. The following should be included:

   almonds (in husk, if possible), olives, coconut (with husk on)
   walnuts (with hulls), cherries
   apricots, peaches, plums
   apples, pears
   strawberries, blackberries
   bananas, cranberries or blueberries, tomatoes
   pumpkin, cucumber, or squash
   orange, lemon, or grapefruit
   pineapple, osage orange, fig

   samaras of maple, elm, and ash
   hazelnuts, acorns
   corn (dry on cob)
   sunflower achenes
   poppy, lily, and other capsules
   mustard, *Lunaria,* and other siliques or silicles
   milkweed, magnolia follicles
   dry bean or pea pods with seeds

2. Spices demonstration; *1 set*

3. Survival plants demonstration, which might include the following:

   dandelion
   dock
   shepherd's purse
   miner's lettuce

   cattails
   bedstraw
   biscuitroot
   lamb's quarters

4. Poisonous plants demonstration, which might include the following:

   pokeweed
   spotted spurge
   Dieffenbachia
   oleander
   yew
   monkshood

   Daphne
   poison ivy/oak
   water hemlock
   buckeye
   poison hemlock
   mayapple

Exercise 20

Ecology (students work in teams)

1. Large balls of string or long metal measuring tapes; *2*
2. Marking pens; *5*

Exercise 21

Genetics (students work in pairs)

1. Ears of hybrid corn to show 9:3:3:1 ratio;* *adequate*
2. Coins

*Instructors should remind students not to pick corn kernels from the cobs.

Laboratory Manual Materials Listing by Carolina Biological Supply Company

Stern's Introductory Plant Biology Laboratory Manual, fourteenth edition, by James E. Bidlack, Published by McGraw-Hill Publishers

To the Instructor: In an effort to make the implementation of learning activities as convenient as possible, McGraw-Hill Publishers and Carolina Biological Supply Company have joined efforts to provide a one-stop source for most of the supplies necessary to complete the experiments in this laboratory manual. All materials may be ordered directly from Carolina Biological Supply Company at (800) 334-5551 or through the company's website at http://www.carolina.com. For more information on the accompanying textbook, *Stern's Introductory Plant Biology,* fourteenth edition, by James E. Bidlack and Shelley H. Jansky, call McGraw-Hill Publishers at (800) 262-4729. Instructors can obtain teaching aids by calling the Customer Service Department at (800) 338-3987, by visiting the McGraw-Hill website at http://www.mhhe.com, or by contacting their local McGraw-Hill sales representative.

Note: Biological supply companies discontinue or change items from time to time. When an item is listed as local, it can usually be obtained from a local store, greenhouse, or herbarium, or another supply company such as Ward's Natural Science ([800 962-2660] or http://www.wardsci.com). Check with your stockroom or purchasing department for additional alternate sources.

Item Name	Catalog Item No.	Quantity per Group	Section
General Materials			
coverslips, glass*	W5-63-2962	see lab	Apparatus
gum eraser	local	1	
medicine dropper, glass*	local	see lab	
microscope slides, glass*	W5-63-2950	see lab	Apparatus
paper, drawing, biology	local	1	
pencil, 3H or harder	local	1	
probe, steel*	W5-62-7400	2	Apparatus
razor blades, single-edged*	local	2	
ruler	W5-70-2608	1	Apparatus
Exercise 1—The Microscope			
lens paper	W5-63-4005	1 per class	Apparatus
microscope, compound*	W5-59-0950	1	Microscopes
microscope, dissecting*	W5-59-1976	1	Microscopes
slides, crossed silk fibers	W5-29-141B	1	Microscope Slides
slides, letter *e*	W5-29-1406	1	Microscope Slides
water, pond*	W5-16-3380	2–3 jars per class	Living Animals
Exercise 2—The Cell			
beaker, 400-ml*	W5-72-1224	1	Apparatus
bowl*	W5-74-1004	4	Apparatus
coverslips, glass*	W5-63-2962	5	Apparatus
*Elodea** sprigs[1]	W5-16-2100	1 sprig	Aquaria & Terraria
foliage plant, fresh**	W5-15-7560	1 flower	Living Plants
IKI solution*	W5-86-9051	2 dropper bottles per class	Chemistry
medicine dropper, glass*	W5-73-6984	1	Apparatus

*Indicates materials that are listed in more than one exercise.

**Request *Tradescantia* when ordering.

[1]Note that in some areas, *Elodea densa* is considered to be an invasive species. In this case, *Elodea canadensis* should be requested.

Item Name	Catalog Item No.	Quantity per Group	Section
microscope, compound*	W5-59-0950	1	Microscopes
microscope slides, glass*	W5-63-2950	5	Apparatus
plant cell model	W5-56-8050	1 per class	Models
potato, white	local	1 per class	
razor blades, single-edged*	local	1	
tomato	local	1 per class	
water, pond*	W5-16-3380	1 bowl per class	Living Animals
water, tap	local	see lab	

Exercise 3—Mitosis

Item Name	Catalog Item No.	Quantity per Group	Section
Allium root tip, mitosis, longisection	W5-30-2396	1	Microscope Slides
modeling clay, blue*	W5-64-4002	1 box per class	Apparatus
modeling clay, green*	W5-64-4000	1 box per class	Apparatus
paper, brown, 60 × 60 cm*	local	1 piece	
plant mitosis (animation)	W5-39-8934	1 per class	CD-ROM
plant mitosis (biophoto sheets, onion root tip)	local	1 pad per class	
plant mitosis set (model)	W5-56-1640	1 set per class	Models

Exercise 4—Roots

Item Name	Catalog Item No.	Quantity per Group	Section
coverslips, glass*	W5-63-2962	1	Apparatus
microscope, compound*	W5-59-0950	1	Microscopes
microscope slides, glass*	W5-63-2950	1	Apparatus
petri dish*	W5-74-1350	1	Apparatus
radish seedlings (seeds)	W5-15-9000	approximately 5	Living Plants
Ranunculus, mature root, cross section	W5-30-2090	1	Microscope Slides
Salix, branch-root origin, cross section	W5-30-1940	1	Microscope Slides
Smilax, mature root, cross section	W5-30-2480	1	Microscope Slides

Exercise 5—Stems

Item Name	Catalog Item No.	Quantity per Group	Section
balsam or Permount in xylene	W5-84-6548	1 dropper bottle per class	Chemistry
basswood twigs, live	local	1	
Begonia stems,* live	W5-15-7260	1	Living Plants
buckeye twigs, dormant	local	1	
clove oil	local	100 ml per class	
Coleus plant*	W5-15-7310	1	Living Plants
coverslips, glass*	W5-63-2962	1	Apparatus
dropper bottle, 30-ml*	W5-71-6552	3 per table	
ethanol 95%,* stock bottle	W5-86-1281	1 liter per class	Chemistry
filter paper, Whatman, 9 cm diam.*	W5-71-2704	1 box per class	Apparatus
Medicago, young & mature stem, c.s.	W5-30-2780	1	
microscope, compound*	W5-59-0950	1	Microscopes
microscope slides, glass*	W5-63-2950	1	Apparatus
razor blades, single-edged*	local	1	
stem type models (dicot)	local	1	
stem type models (monocot)	W5-56-8770	1	Models
stem type review sheets (herbaceous dicot)	local	1 pad per class	
stem type review sheets (monocot)	local	1 pad per class	
stem type review sheets (monocot & dicot)	local	1 pad per class	
stem type review sheets (woody dicot)	local	1 pad per class	
Tilia, 2-yr. stem, cross section	W5-30-2816	1	Microscope Slides
watch glass	W5-74-2356	1	Apparatus
water	local	see lab	
xylene	W5-89-8741	1 dropper bottle per class	Chemistry
Zea mays, stem, cross section	W5-30-3296	1	Microscope Slides

*Indicates materials that are listed in more than one exercise.

Item Name	Catalog Item No.	Quantity per Group	Section
Each student should have 1 set of dropper bottles of:			
Eosin solution:			
eosin	local	500 mg	
ethanol, 95%*	W5-86-1281	100 ml	Chemistry
Gentian violet stain:			
ethanol, 95%*	W5-86-1281	20 ml	Chemistry
formalin, 5%*	W5-86-3531	80 ml	Chemistry
gentian violet*	local	1 g	
Phoroglucinol stain:			
phoroglucinol	local	1 gm	
ethanol, 95%*	W5-86-1281	100 ml	Chemistry

Exercise 6–Leaves

Item Name	Catalog Item No.	Quantity per Group	Section
insectivorous plant collection*	W5-15-7185	1 per class	Living Plants
leaves, review sheets	local	1 set per class	
microscope, compound*	W5-59-0950	1	Microscopes
monocot leaf model	W5-56-8801	1 per class	Models
Pinus, leaf cross section	W5-30-1376	1	Microscope Slides
Sedum plant	local	1	
Syringa, leaf cross section	W5-30-3790	1	Microscope Slides
twigs w/ compound leaves	local	1 set	
Zebrina plant	W5-15-7590	1	Living Plants

Exercise 7–Plant Propagation

Item Name	Catalog Item No.	Quantity per Group	Section
African violet plant*	W5-15-7343	10 leaves	Living Plants
aluminum foil*	W5-71-3210	1 roll per class	Apparatus
apples, ripe	local	1	
*Begonia** stem, 1.5 dm long	W5-15-7260	1 stem	Living Plants
Bunsen burner, artificial* **or**	W5-70-6706 or 6709	1	Apparatus
Bunsen burner, natural* gas	W5-70-6706		Apparatus
carrots	local	1	
containers for plant bulbs	W5-71-6061	1	Apparatus
fern allies set (*Equisetum* stems)	W5-15-6951	1	Living Plants
filter paper*	W5-71-2734	1 box per class	Apparatus
flats, plastic	W5-66-5918	4	Living Plants
forceps, 10-in.	W5-62-4335	1	Apparatus
grafting wax, sealing compound	local	see lab	
jar, wide-mouth, glass*	W5-71-5271	2	Apparatus
knife	local	1	
light banks (for propagated materials in lab)	W5-15-8998	1 per class	Living Plants
Lily bulbs (for lily plant)	local	1	
liquid bleach solution,* 10%	W5-88-9640	500 ml per class	Chemistry
matches*	W5-97-2148	1 box per class	
melon-ball scoop	local	see lab	
nutrient agar slants	W5-82-6102	1	Microbiological Media
pan* (for African violet)	W5-62-9380	1	Apparatus
peanuts, raw	local	1/4 lb.	
petri dish*	W5-74-1350	2	Apparatus
pineapple, whole	local	1	
plastic bags	W5-71-3091	1	Apparatus
potatoes, sweet	local	1/2 potato	
potatoes, white	local	1/2 potato	
pots, plastic*	W5-66-5776	2	Living Plants
potting soil*	W5-15-9705	1 bag per class	Living Plants

*Indicates materials that are listed in more than one exercise.

Item Name	Catalog Item No.	Quantity per Group	Section
probe, steel*	W5-62-7400	1	Apparatus
razor blades, single-edged*	local	1	
Rootone	W5-20-7764	1 per class	Physiology
rose cuttings	W5-19-1194	see lab	
rubber bands	W5-97-3029	2	
rubber strips, flat or plastic tape	W5-19-9278A	see lab	
test-tube rack	W5-73-1876	1	Apparatus
tissues, facial	local	2 boxes per class	
toothpicks*	local	1 box per class	
vermiculite, 1 peck	W5-15-9721	l bag per class	Living Plants
wax paper	local	1 roll per class	
willow or poplar twigs, dormant, 0.75 cm diam.	local	1	
willow or poplar twigs, dormant, 1.5 cm diam.	local	1	

Exercise 8—Cell Components and Products

Item Name	Catalog Item No.	Quantity per Group	Section
beaker, 50-ml	W5-72-1220	2 per table	Apparatus
beaker, 400-ml*	W5-72-1224	1 per table	Apparatus
Benedict's solution	W5-84-7111	5 ml	Chemistry
Biuret reagent	W5-84-8211	120 ml per class	Chemistry
bottle, 250-ml	W5-71-6314	8	Apparatus
Brazil nut	local	5 per class	
Bunsen burner, artificial* **or**	W5-70-6706 or 6709	1	Apparatus
Bunsen burner, natural* gas	W5-70-6706		Apparatus
cabbage, purple or red	local	1 per class	
dishpan*	W5-62-9380	1	Apparatus
filter paper, Whatman 9 cm diam.*	W5-71-2734	see lab	Apparatus
forceps, fine*	W5-62-4024	1	Apparatus
glucose solution:		250 ml per class	
glucose*	W5-85-7430	5 g per class	Chemistry
water, distilled*	W5-19-8697	1 liter per class	Plant Tissue Culture
hydrochloric acid, 0.1 M	W5-86-7821	10 ml per table	
ice cubes	local	1 tray	
IKI solution* (iodine reagent)	W5-86-9051	30 ml per class	Chemistry
insulated wire gauze* (for hot-water bath)	local	1	
matches*	W5-97-2148	1 book per class	
medicine dropper, glass*	W5-73-6902	1	Apparatus
mesh, fiberglass, 4-in. diameter	W5-97-2831	1	
microscope, compound*	W5-59-0950	1	Microscopes
microscope slides, glass*	W5-63-2950	1	Apparatus
mint leaves	local	20	
mortar and pestle, 60-ml	W5-74-2890	1 per table	Apparatus
petri dish*	W5-74-1350	1 per table	Apparatus
protein solution:		250 ml per class	
albumen, powdered	W5-84-2250	10 g per class	Chemistry
water, distilled*	W5-19-8697	1 liter per class	Plant Tissue Culture
sodium hydroxide, 0.1 M	W5-88-9551	10 ml per table	Chemistry
starch solution:		250 ml per class	
soluble starch*	W5-89-2530	1 g per class	Chemistry
water, boiling	local	100 ml per class	
water, distilled*	W5-19-8697	1 liter per class	Plant Tissue Culture
Sudan IV stain	W5-89-2993	1 dropper bottle per class	Chemistry
Tes-tape	W5-89-3840	1 roll per class	Chemistry
test tube, 13 × 100 mm*	W5-73-1008	9	Apparatus

*Indicates materials that are listed in more than one exercise.

Item Name	Catalog Item No.	Quantity per Group	Section
test-tube brush*	W5-70-6030	1 per class	Apparatus
test-tube holder*	W5-70-2900	1	Apparatus
test-tube rack*	W5-73-1871	1	Apparatus
tissues, facial*	local	see lab	
tripod*	W5-70-6950	1	Apparatus
watch glass (lid)	W5-74-2356	1	Apparatus

Exercise 9–Diffusion, Growth, and Hormones

Item Name	Catalog Item No.	Quantity per Group	Section
agar in petri dishes for crystal dye diffusion	W5-82-1862	1	Microbiological Media
Balsam plant (*Impatiens* seed)	local	1	
beaker, 600-ml	W5-72-1225	2	Apparatus
bowl*	W5-74-1004	1	Apparatus
Coleus plant*	W5-15-7312	4	Living Plants
coverslips, glass*	W5-63-2962	3	Apparatus
dropper bottle, 30-ml*	W5-71-6552	3	
dwarf corn (seed) plant	W5-17-7110	3 plants	Genetics
Elodea plant in water	W5-16-2100	4 sprigs per class	Aquaria & Terraria
food coloring (aqueous eosin)	W5-89-8030	500 ml per class	Chemistry
gibberellic acid solution	W5-19-8633	.1 microgram per class	Plant Tissue Culture
grid marker	local	1	
IAA (auxin)	W5-20-7603	1 small jar per class	Physiology
IKI solution*	W5-86-9051	2 dropper bottles	Chemistry
India ink	local	2 dropper bottles	
ink pad	local	1	
lanolin	W5-87-1810	1 small jar per class	Chemistry
matches*	W5-97-2148	1 box per class	
microscope, compound*	W5-59-0950	1	Microscopes
microscope slides, glass*	W5-63-2950	3	Apparatus
paper towels	W5-63-3954	2	
plant tags, tie-on	W5-65-7556	2	Apparatus
potassium permanganate crystals	W5-88-4130	several	Chemistry
potting soil and pots	local	2 pots with soil per table	
razor blades, single-edged*	local	1	
Rootone rooting compound*	W5-20-7764	1 container per class	Physiology
scissors*	W5-64-4772	1	Apparatus
sucrose solution, 25%*	W5-89-2860	2 dropper bottles	Chemistry
toothpicks*	local	see lab	
wetting agent	W5-20-7861	1 small jar per class	Physiology
Dialysis demo: (1 per class)			
beaker, 100-ml*	W5-72-1221	1 per class	Apparatus
dialysis tubing, presoaked	W5-68-4212	1 dm per class	Apparatus
dialyzing solution (glucose)*	W5-85-7430	20 g per class	Chemistry
dialyzing solution (soluble starch)*	W5-89-2530	1 g per class	Chemistry
string, starch-free ball of*	W5-97-2222	small ball	
Diffusion demo: (1 per class)			
ammonium hydroxide	W5-84-4033	50 ml per class	Chemistry
buret, glass	W5-73-8091	1 per class	Apparatus
clamp	W5-70-7362	1 per class	Apparatus
Erlenmeyer flask, 125-ml	W5-72-6670	1 per class	Apparatus
litmus paper, red	W5-89-5510	1 per class	Chemistry
meterstick	W5-70-2620	1 per class	Apparatus
ring stand	W5-70-7161	1 per class	Apparatus
stopper* for buret	W5-71-2400	1 per class	Apparatus

*Indicates materials that are listed in more than one exercise.

Item Name	Catalog Item No.	Quantity per Group	Section
Ethylene gas demo: (1 per class)			
apple, ripe	local	1 per class	
beaker, 100-ml*	W5-72-1221	2 per class	Apparatus
bell jar*	local	2 per class	
holly sprigs	local	2 per class	
Etiolation demo: (1 per class)			
bean seeds	W5-15-8335	10 seedlings per class	Living Plants
corn seeds*	W5-15-9283	10 seedlings per class	Living Plants
pea seeds	W5-15-8883	10 seedlings per class	Living Plants
pots, plastic*	W5-66-5776	6 per class	Living Plants
potting soil*	W5-15-9705	half bag per class	Living Plants
Gravitropism demo: (1 per class)			
Coleus plant*	W5-15-7312	1 per class	Living Plants
Osmosis demo: (1 per class)			
osmometer	W5-68-4100	3 per class	Apparatus
water	local	see lab	
Phototropism demo: (1 per class)			
box (to fit over plant)	local	1 per class	
Coleus plant*	W5-15-7312	1 per class	Living Plants
Transpiration demo: (1 per class)			
bell jar*	local	1 per class	
Coleus plant*	W5-15-7312	2 per class	Living Plants
Turgor movements demo: (1 per class)			
insectivorous plants set*	W5-15-7185	1 set per class	Living Plants
Mimosa pudica	W5-15-7520	1 set per class	Living Plants

Exercise 10–Photosynthesis

Item Name	Catalog Item No.	Quantity per Group	Section
acetone (chromatography solvent)	W5-84-1500	10 ml per class	Chemistry
aluminum foil*	W5-71-3210	1 medium sheet	Apparatus
beaker, 400-ml*	W5-72-1224	4	Apparatus
black paper	W5-97-1266	1 pack per class	
Bunsen burner, artificial* **or**	W5-70-6706 or 6709	1 per table	Apparatus
Bunsen burner, natural* gas	W5-70-6706		Apparatus
camel's-hair brush	local	1	
chromatography strips, 1 × 15 cm	local	25 per class	
Coleus plants*	W5-15-7310	2 plants per class	Living Plants
concentrated alcoholic extract of chlorophyll	local	400 ml	
dishpan*	W5-62-9380	1 per class	Apparatus
dropper bottle, 30-ml*	W5-71-6552	1	Apparatus
Elodea sprigs*	W5-16-2100	6 sprigs per class	Aquaria & Terraria
ethanol, 95%*	W5-86-1281	1 liter	Chemistry
forceps, fine*	W5-62-4024	1	Apparatus
glass tubing* (size of soda straws)	W5-71-1145	1	Apparatus
green plant pigment, fresh	local	1 dropper bottle	
IKI solution* (iodine reagent)	W5-86-9051	1 dropper bottle	Chemistry
insulated wire gauze* (for hot-water bath)	W5-70-6902	1	Apparatus
petri dish*	W5-74-1350	2	Apparatus
petroleum ether* (chromatography solvent)	W5-87-9540	90 ml per class	Chemistry
phenol red solution	W5-87-9873	1 dropper bottle	Chemistry
stopper* (for test tube)	W5-71-2400	2	Apparatus
test tube, 13 × 100 mm*	W5-73-1008	5	Apparatus
test-tube brush*	W5-70-6030	1 per class	Apparatus
test-tube holder*	W5-70-2900	1 per table	Apparatus
test-tube rack*	W5-73-1871	1	Apparatus
tripod*	W5-70-6950	1 per table	Apparatus

*Indicates materials that are listed in more than one exercise.

Item Name	Catalog Item No.	Quantity per Group	Section
Plant pigment extract: (see Appendix)			
acetone	W5-84-1481	100 ml	Chemistry
blender	local	1 per class	
cheesecloth	W5-71-2690	1 per class	Apparatus
glass wool	W5-71-3000	1 per class	Apparatus
ivy leaves	W5-15-7390C	see lab	Living Plants
petroleum ether*	W5-87-9540	15 ml per class	Chemistry
separatory funnel, 250-ml	local	1 per class	
vials with caps	W5-71-5004	3 per class	Apparatus
water, distilled*	W5-19-8697	20 ml per class	Plant Tissue Culture

Exercise 11—Water in Plants; Respiration; Digestion

Item Name	Catalog Item No.	Quantity per Group	Section
apple	local	1	
barium hydroxide (baryta water)	W5-84-6818	1 liter	Chemistry
barley seedlings (seeds)	W5-15-9223	20	Living Plants
beaker, 400-ml*	W5-72-1224	1	Apparatus
bean sprouts	local	10	
bottle, wide-mouth	W5-19-9253	2	Apparatus
bottles, insulated (vacuum bottle)	local	2	
bottles of food coloring	local	3 per table	
Bunsen burner, artificial* **or**	W5-70-6706 or 6709	1	Apparatus
Bunsen burner, natural* gas	W5-70-6706		Apparatus
carnations	local	3 per table	
corn kernels,* viable	W5-15-9283	30	Living Plants
Diastase	W5-85-7538	2 to 3 drops	Chemistry
electric fan, small	W5-70-1275	1	Apparatus
formalin solution,* 5%	W5-86-3531	1 liter per class	Chemistry
hydrogen peroxide	W5-86-8063	1 dropper bottle	Chemistry
jar, wide-mouth, glass*	W5-71-5261	1	Apparatus
liquid bleach solution,* 10%	W5-88-9640	1 pint	Chemistry
matches*	W5-97-2148	1 box per class	
metal gauze*	W5-70-6902	1 square/tripod	Apparatus
microscope, compound*	W5-59-0950	1	Microscopes
microscope slides, glass*	W5-63-2950	3	Apparatus
potato	local	1	
potometer (Transpiration Kit)	W5-20-6010	1 kit per class	Physiology
starch powder*	W5-89-2530	see lab	Chemistry
stoppers to fit around thermometers	W5-71-2438	2	Apparatus
stopper with 1 hole	W5-71-2440	1	Apparatus
stopper with 2 holes	W5-71-2470	1	Apparatus
tetrazolium (1% aqueous solution)	W5-89-6930	1 dropper bottle	Chemistry
thermometer	W5-74-5422	2	Apparatus
tripod*	W5-70-6950	1 per table	Apparatus
tubing, glass*	W5-71-1145	2 pieces	Apparatus
tubing, rubber	W5-71-1404	1 piece	Apparatus
twig, woody with broad leaves	local	1	
watch glass, Syracuse	W5-74-2320	1	Apparatus
yeast suspension in 10% sucrose solution		1 dropper bottle	
sucrose*	W5-89-2860	10 g	Chemistry
yeast	W5-17-3234	1 g	Genetics

Exercise 12—Meiosis and Alternation of Generations

Item Name	Catalog Item No.	Quantity per Group	Section
meiosis, bioreview sheets	local	1 pad per class	
modeling clay, blue*	W5-64-4002	1 box per class	Apparatus

*Indicates materials that are listed in more than one exercise.

Item Name	Catalog Item No.	Quantity per Group	Section
modeling clay, green*	W5-64-4000	1 box per class	Apparatus
paper, brown,* 60 × 60 cm	local	1 piece	

Exercise 13—Domain and Kingdom Survey

Item Name	Catalog Item No.	Quantity per Group	Section
agar plate with bacterial colonies: *B. cereus*	W5-15-4869	1	Living Plants
agar plate with bacterial colonies: *M. luteus*	W5-15-5156	1	Living Plants
cyanobacteria, agar slant: *Oscillatoria*	W5-15-1865	1	Living Plants
bracket fungus, preserved*	local	1 set	
Chlorella, green algae	W5-15-2069	1 culture	Living Plants
club mosses (*Lycopodium* set)	local	1 set	
corn smut, preserved*	W5-29-8266	1 set	Microscope Slides
Cosmarium, green algae	W5-15-2140	1 culture	Living Plants
Dictyostelium, slime mold	W5-15-5995	1	Living Plants
Didymium nigripes, slime mold	W5-15-6002	1	Living Plants
dinoflagellates: *Peridinium**	W5-15-3290	1 culture or portion	Living Plants
Euglena culture	W5-15-2800	1	Living Plants
morel	local	1	
mushrooms	local	1 set	
Pediastrum, green algae	W5-15-2430	1 culture	Living Plants
Physarum, slime mold*	W5-15-6190	1	Living Plants
rust, preserved*	local	1 set	
Spirogyra, green algae*	W5-15-2525	1 culture	Living Plants
Stemonitis, slime mold	W5-15-6002	1	Living Plants
stinkhorn*	local	1 set	
true fungi including			
truffles	local	1 set	
Volvox, green algae	W5-15-2655	1 culture	Living Plants
herbarium sheets:			
Marine Algae Set of 12 (includes the following):	local	1 set	
Codium	local	1	
Desmarestia	local	1	
Gelidium	local	1	
Laminaria	local	1	
Nereocystis	local	1	
Porphyra	local	1	
Postelsia	local	1	
Ulva	local	1	
4 more of teacher's choice	local	4	
living specimen of:			
conifers	W5-15-7064	1 set	Living Plants
cycad*	local	1 set	
Equisetum & Selaginella	W5-15-6950	1 set	Living Plants
ferns*	W5-15-6842	1 set	Living Plants
flowering plants (African violet)*	W5-15-7343	1 set	Living Plants
flowering plants (*Begonia*)*	W5-15-7260	1 set	Living Plants
Ginkgo	local	1 set	
liverworts	W5-15-6540	1 set	Living Plants
mosses (w/conspicuous sporophytes)*	W5-15-6695	1 set	Living Plants
Psilotum	W5-15-7000	1 set	Living Plants

Exercise 14—Domains (Kingdoms) Archaea and Bacteria; Kingdom Protista

Item Name	Catalog Item No.	Quantity per Group	Section
Anabaena, live culture	W5-15-1710	1 culture per class	Living Plants
bacteria plates, gram-neg.: *S. marcescens*	W5-15-5451	1 plate per class	Living Plants
bacteria plates, gram-pos.: *Sarcina lutea*	W5-15-5421	1 plate per class	Living Plants

*Indicates materials that are listed in more than one exercise.

Item Name	Catalog Item No.	Quantity per Group	Section
bacterial plates with variety of colonies	W5-15-4756	1 set per class	Living Plants
bread, white	local	1 loaf per class	
Bunsen burner, artificial* **or**	W5-70-6706 or 6709	1	Apparatus
Bunsen burner, natural* gas	W5-70-6706		Apparatus
Chlamydomonas live culture	W5-13-1738	1 culture per class	Living Animals
Codium preserved specimen	local	1 per class	
Costaria herbarium specimen	local	1 per class	
coverslips, glass*	W5-63-2962	12	Apparatus
Desmarestia herbarium specimen	local	1 per class	
diatomaceous earth	local	1 oz. per class	
dinoflagellates: *Amphidinium*	W5-15-3240	small amount	Living Plants
dinoflagellates: *Peridinium**	W5-15-3290	small amount	Living Plants
ethanol, 95%*	W5-86-1281	120 ml	Chemistry
Euglena live culture	W5-13-1768	1 culture per class	Living Animals
Formalin, 5%*	W5-86-3531	80 ml	Chemistry
Gelidium herbarium specimen	local	1 per class	
gentian (crystal) violet*	local	1 g	
Gigartina herbarium specimen	local	1 per class	
Gram's iodine reagent:			
iodine	W5-86-8970	0.5 g per class	Chemistry
potassium iodide	W5-88-3790	1 g per class	Chemistry
Laminaria herbarium specimen	local	1 per class	
microscope, compound*	W5-59-0950	1	Microscopes
microscope, dissecting*	W5-59-1822	1	Microscopes
microscope slides, glass*	W5-63-2950	see lab	Apparatus
Nereocystis herbarium specimen	local	1 per class	
Oedogonium, whole-mount slide	W5-29-6500	1	Microscope Slides
Oedogonium live culture	W5-15-2400	1 culture per class	Living Plants
petri dish, small	W5-74-1346	1	Apparatus
Porphyra herbarium specimen	local	1 per class	
Postelsia herbarium specimen	local	1 per class	
safranin O dye:			
ethanol, 95%*	W5-86-1281	100 ml per class	Chemistry
safranin O	W5-88-7039	1 g per class	Chemistry
Scenedesmus live culture	W5-15-2510	1 culture per class	Living Plants
slime mold plasmodia, live	W5-15-6193	1 per class	Living Plants
slime mold set, preserved	local	1 set per class	
slime mold sporangia, expendable & non*	W5-15-6190	1 per class	Living Plants
Spirogyra, whole-mount slide	W5-29-6548	1	Microscope Slides
Spirogyra live culture*	W5-15-2525	1 culture per class	Living Plants
Ulothrix live culture	W5-15-2640	1 culture per class	Living Animals
Ulothrix, whole-mount slide*	W5-29-6602	1	Microscope Slides
Ulva herbarium specimen	local	1 per class	
Volvox live culture	W5-13-1860	1 culture per class	Living Animals
water, pond*	W5-16-3380	1 bowl per class	Living Animals

Exercise 15–Kingdom Fungi (Mycota)

Item Name	Catalog Item No.	Quantity per Group	Section
bird's-nest fungi, preserved	local	1	
bracket fungus, preserved*	local	1	
Coprinus, pileus slide	W5-29-8176	1	Microscope Slides
Coprinus, preserved	local	1	
corn smut, preserved*	W5-29-8266	1	Microscope Slides
crustose, foliose, fruticose lichen (set)	W5-15-6400	1 set per class	Living Plants
earth stars, preserved	local	1	
microscope, compound*	W5-59-0950	1	Microscopes

*Indicates materials that are listed in more than one exercise.

Item Name	Catalog Item No.	Quantity per Group	Section
Penicillium, whole-mount slide	W5-29-7968	1	Microscope Slides
Peziza, apothecia, cross-section slide	W5-29-7980	1	Microscope Slides
Physcia, thallus slide	W5-29-8476	1	Microscope Slides
puffballs, preserved	local	1	
Rhizopus, living sample	W5-15-6222	1	Microbiology
Rhizopus, whole-mount slide	W5-29-7770	1	Microscope Slides
rust, preserved*	local	1	
stinkhorn*	local	1	

Exercise 16–Kingdom Plantae: Bryophytes and Ferns

Item Name	Catalog Item No.	Quantity per Group	Section
fern plants*	W5-15-6842	1	Living Plants
fern plants (1 with expendable fronds)	W5-15-6902	1	Living Plants
hornworts, live (call for availability)	W5-16-2041	1	Living Plants
Marchantia, live (antheridia)	W5-15-6544	1	Living Plants
Marchantia, live (archegonia)	W5-15-6546	1	Living Plants
microscope, compound*	W5-59-0950	1	Microscopes
Mnium, archegonial, longisection slide	W5-29-9028	1	Microscope Slides
mosses, live with sporophytes attached*	W5-15-6695	1	Living Plants
moss protonema slide	W5-29-8900	1	Microscope Slides
prothalli, bisexual, whole-mount slide	W5-29-9290	1	Microscope Slides
prothalli, live	W5-15-6879	1	Living Plants
protonemata, living	W5-15-6681	1	Living Plants

Exercise 17–Kingdom Plantae: Gymnosperms

Item Name	Catalog Item No.	Quantity per Group	Section
conifer pollen	local	small amount	
cycads specimen, living*	local	1 per class	
Ephedra Life Cycle Set (preserved)	local	1 per class	
Ginkgo Life Cycle Set (preserved)	local	1 per class	
Gnetum & Welwitschia specimens, pictures of	W5-45-8004	1 per class	Books
microscope, compound*	W5-59-0950	1	Microscopes
Pinus, archegonium, longisection slide	W5-30-1478	1	Microscope Slides
Pinus female (seed) cones	local	1	
Pinus male (pollen) cones, fresh	local	1	

Exercise 18–Kingdom Plantae: Angiosperms (Flowering Plants–Phylum Magnoliophyta)

Item Name	Catalog Item No.	Quantity per Group	Section
flower, model of	local	1 per class	
flowers, large, fresh with superior ovaries	local	see lab	
flowers, live on display	local	see lab	
Lilium, embyo sac, cross-section slide	W5-30-4688	1	Microscope Slides
Lilium, mature anthers, cross-section slide*	W5-30-4586	1	Microscope Slides
microscope, dissecting*	W5-59-1976	1	Microscopes

Exercise 19–Fruits, Spices, and Survival Plants

Item Name	Catalog Item No.	Quantity per Group	Section
Fruit Types Set	local	1 set per class	
Medicinal Plants Set (Herbarium Specimens)	local	1 set per class	
Poisonous Plants Set (Herbarium Specimens)	local	1 set per class	
spices, various	W5-15-9530	1 set per class	Living Plants

Exercise 20–Ecology

Item Name	Catalog Item No.	Quantity per Group	Section
marking pen	local	1	
microscope, compound*	W5-59-0950	1	Microscopes
string, ball of*	W5-97-2222	see lab	

Exercise 21–Genetics

Item Name	Catalog Item No.	Quantity per Group	Section
coins	local	2	
F_2 generation ears of corn (purple & yellow)	W5-17-6600	1	Genetics

*Indicates materials that are listed in more than one exercise.